Cryptids

The Fascinating World of Unknown Creatures

(A Deep Dive Into the Cryptids, Folklore, and Hidden Stories From Coast)

James Bell

I0089825

Published By **Barry Ackles**

James Bell

Cryptids: The Fascinating World of Unknown Creatures (A Deep Dive Into the Cryptids, Folklore, and Hidden Stories From Coast)

ISBN 978-1-7774070-2-5

No part of this guidebook shall be reproduced in any form without permission in writing from the publisher except in the case of brief quotations embodied in critical articles or reviews.

Legal & Disclaimer

The information contained in this book is not designed to replace or take the place of any form of medicine or professional medical advice. The information in this book has been provided for educational & entertainment purposes only.

The information contained in this book has been compiled from sources deemed reliable, and it is accurate to the best of the Author's knowledge; however, the Author cannot guarantee its accuracy and validity and cannot be held liable for any errors or omissions. Changes are periodically made to this book. You must consult your doctor or get professional medical advice before using any of the suggested remedies, techniques, or information in this book.

Table Of Contents

Chapter 1: Wolf Woman of Mobile 1

Chapter 2: Deciphering the Boggy Creek Enigma 14

Chapter 3: Selbyville Swamp Monster 28

Chapter 4: Van Meter Monster 53

Chapter 5: Dover Demon 78

Chapter 6: Alkali Lake Monster 104

Chapter 7: The First Encounter 116

Chapter 8: Loveland Frogmen 137

Chapter 9: Stories 144

Chapter 1: Wolf Woman of Mobile

Mobile, Alabama, has prolonged been referred to as the birthplace of an enigmatic legend – a hybrid creature called the Wolf Woman. This half of of-wolf, half of of-woman cryptic reportedly haunted the city and its surrounding areas sooner or later of April 1971, sending waves of worry and interest amongst its residents.

This tale unfolds in the dense swamps and marshlands of Mobile, in which numerous locals declare to have encountered the mystifying creature, described as bearing the pinnacle torso of a strikingly adorable girl and the decrease quarters of a wild wolf. Witnesses stated the creature transferring with the speed and agility of a wolf, traversing on all fours thru the murky shadows of the night time time. With a propensity for roaming the areas of

Plateau and Port City, the Wolf Woman have become the problem of severa sightings that harassed the Mobile community.

On April eighth, 1971, the neighborhood newspaper, The Mobile Register, ran a story on the unusual sightings, whole with an example of the cryptid as envisaged through their illustrator. In the instances that determined, the newspaper grows to be inundated with reviews from locals who had visible or heard of the mysterious creature.

As is common in folklore, legends of half of of-human, half-animal beings have always captivated human imagination, forming the spine of numerous mythologies across numerous cultures. The werewolf, as an instance, is a famous entity in European and Native American folklore and has parallels with the skinwalker, a form-transferring creature located in Native

American way of life. The skinwalker, in line with legend, can rework from human to animal, regularly the usage of malevolent manner to accumulate this metamorphosis. Once transformed, the creature is said to wreak havoc and instill fear maximum of the neighborhood populace.

In the case of the Wolf Woman of Mobile, a few speculate that she can also additionally have been a kind of skinwalker. From the time of the primary mentioned sighting in early April 1971, descriptions of the Wolf Woman continuously depicted her with a lady's better body and the lower limbs of a wolf.

Witnesses defined their encounters with the Wolf Woman in numerous tactics, with one person recalling her as "a girl and a wolf, quite and hairy." The Mobile Register's smartphone traces had been flooded with over fifty calls from locals claiming to have seen or been chased with the useful resource of manner of the creature. Some even said sightings of the Wolf Woman in their backyards. An unnerving rumor circulated, suggesting that the Wolf Woman have become a former circus sideshow enchantment who had with the aid of a few method managed to break out.

Amidst all the chaos and panic, close by regulation enforcement have emerge as forced to investigate the sightings, treating them as credible reviews of a functionality wild animal at the unfastened. Thankfully, irrespective of the priority and paranoia that gripped the city, there have been no recorded times of the Wolf Woman inflicting physical harm to all people.

Ultimately, the fact behind the Wolf Woman of Mobile stays shrouded in mystery, just like the shadows of the murky swamps from wherein she became believed to have emerged. Whether she modified right into a figment of collective imagination, a misidentified animal, or virtually a cryptid of mythological proportions, the legend of the Wolf Woman remains an interesting bankruptcy in the rich tapestry of American folklore.

Alaska: Tizheruk

The Tizheruk is a captivating and mysterious cryptid believed to inhabit the frigid waters surrounding Alaska, specially near King and Nunivak Islands in the Bering Sea. Described as a sea serpent-like creature, the Tizheruk is characterized via manner of its seven-foot-lengthy head, a flipper tail, and an everyday length of as a minimum fifteen feet.

The Tizheruk is an ambush predator that generally resides in rocky streams, the usage of its specific physical skills to mixture in with its environment and strike unsuspecting prey. Its pores and pores and skin is translucent, revealing its internal organs, which, while submerged, resemble rocks, therefore supplying the creature with an powerful camouflage. The enamel of the Tizheruk resemble those of deep-sea fish, further enhancing its predatory prowess.

Over the years, there have been severa claims of Tizheruk sightings, with some even reporting the invention of mysterious carcasses washed ashore, believed to be remnants of this elusive creature. However, upon further examination, the ones carcasses were usually identified as stays of appeared marine species, which includes whales and sharks.

The Tizheruk has regularly been compared to at least one-of-a-type cryptids and marine animals, which includes the Wasgo, Sisiutl, Steller's Sea Ape, and Cadborosaurus. Some cryptozoologists have even proposed that the Tizheruk can be a form of pinniped, carefully associated with the Antarctic leopard seal (Hydrurga leptonyx), therefore serving as its northern hemisphere counterpart. The bodily similarities among the Tizheruk and the leopard seal similarly assist this concept.

Historically, the Tizheruk has been an crucial a part of network folklore, with ethnologist John White documenting money owed from King Island population. These locals had a strict taboo regarding the Tizheruk, sharing memories of hunters who had fallen victim to this ambitious predator. The islanders also associated the Tizheruk with non secular and mythical significance, fearing its sturdy presence.

Mackal, throughout his research in 1983, determined that the Coast Guard personnel stationed nearby have been aware of the creature, noting its uncommon look inside the place. The locals, referred to as "Eskimos" via Mackal, regarded the Tizheruk with a mixture of reverence and worry, associating it with diverse myths and legends.

The islanders said that the Tizheruk regularly revealed first-rate its head, neck, and sometimes its tail at the identical time as the relaxation of its frame remained submerged. When developing from the water, the creature have emerge as said to stand seven to 8 ft tall, with a snake-like head and a flippered tail. The locals believed that they may appeal to the Tizheruk via tapping at the insides in their boats, a conduct that has additionally been determined in leopard seals.

While the exact nature of the Tizheruk stays shrouded in thriller, its presence continues to be a supply of fascination and intrigue, shooting the imaginations of each locals and cryptozoologists alike. Whether it is a however-to-be-determined species, a very unique form of pinniped, or virtually a figment of folklore, the Tizheruk stays an crucial a part of Alaska's wealthy tapestry of myths and legends.

Arizona: Mogollon Monster

The Mogollon Monster, additionally called the Arizona Bigfoot, is an enigmatic creature stated to inhabit the primary and japanese components of Arizona alongside the Mogollon Rim. This beast stands at over seven feet tall, with descriptions akin to Bigfoot. It has large, pink eyes and a frame protected in lengthy black or reddish-brown hair, with its chest, face, palms, and ft left naked. A stinky scent, corresponding to a mixture of vain fish, skunk, decaying peat moss, and the musk of a snapping turtle, is regularly associated with sightings of this cryptid.

This nocturnal, omnivorous creature is referred to to be territorial and on occasion violent. Witnesses have defined it walking with extensive, inhuman strides, leaving in the back of footprints measuring up to 22 inches. It's said to mimic the sounds of numerous herbal world, emit uncommon whistle sounds, find out campsites throughout the night time time, or even construct nests from pine needles, twigs, and leaves. There are also reviews of the Mogollon Monster throwing stones from hidden locations and decapitating deer and specific natural international in advance than consumption.

The earliest regarded sighting of this creature end up documented in a 1903 model of The Arizona Republican. I.W. Stevens said seeing a creature close to the Grand Canyon with "prolonged white hair and a matted beard that reached its knees." It had talon-like arms with claws at the least inches extended and emitted "the wildest, most unearthly screech."

Over the years, severa exceptional sightings and encounters had been said. For example, in the mid-1940s, cryptozoologist Don Davis stated a Boy Scout journey near Payson, Arizona, in which he noticed a massive creature with a very square face and huge fingers and chest. The creature become blanketed in hair, except for its face, which appeared inside the critical with out hair. Marjorie Grimes, a resident of Whiteriver, Arizona, moreover said a couple of sightings of the creature from 1982 to 2004, describing it as black, tall, and on foot with large strides.

While a few locals staunchly agree with in the lifestyles of the Mogollon Monster, mainstream biologists and scientists are skeptical. Professor Stan Lindstedt from Northern Arizona University labeled the creature as a part of mythology in preference to technology. Despite the numerous anecdotal reviews, no concrete proof has been determined to affirm the lifestyles of the Mogollon Monster, and some attribute the sightings to hoaxes or misidentifications of different massive mammals which embody black bears, mountain lions, or elk.

In Arizona folklore, the Mogollon Monster has come to be an iconic decide with diverse memories and legends woven spherical its life. These memories variety from pioneers being attacked through the creature, Native American chiefs remodeling to scare away their clans, to a tormented Native American seeking out revenge. The monster has additionally determined its manner into network colour publications, works of fiction, and even campaigns in competition to littering, with Dolan Ellis' music "Mogollon Monster" being used to discourage littering inside the desolate tract areas of Arizona.

Arkansas: Fouke Monster

Nestled deep within the rugged landscapes of Arkansas is a tale that has enthralled and puzzled citizens and location visitors for generations. The Fouke Monster, often called the Boggy Creek Monster, is an eerie determine believed to haunt the environment of the small metropolis of Fouke. Originating from reminiscences as early because of the reality the 1800s, this cryptid is described as a frightening creature reputation among 7 to eight toes tall. Strikingly humanoid in stature however beastly in look, its form is covered in a rough coat of hair, important many to liken it to primates.

Chapter 2: Deciphering the Boggy Creek Enigma

The Boggy Creek Monster has firmly rooted itself inside the heart of Fouke, a old fashioned city located a stone's throw from Little Rock, Arkansas. Legends pinpoint Fouke due to the truth the debut vicinity for sightings of this mystifying cryptid, with descriptions detailing a tall, hulking creature included in thick, coarse hair.

Documented sightings trace back to the early 1800s, culminating in the 1900s whilst Fouke became a hotspot for encounters. Some attribute the numerous sightings to the beast's nocturnal behavior. Nevertheless, sunlight hours sightings, consisting of the exceptional within the Sulfur River Wildlife Area in 2000, challenge this notion.

The monster's impact on famous lifestyle is plain in the more than one movies it stimulated, with "The Legend of Boggy Creek" from 1973 taking middle level. This film now not simplest spread the legend national but moreover substantially impacted the neighborhood financial device. The 1971 attack at the Ford residence stands out as one of the maximum harrowing episodes associated with the beast. Accounts detail the creature's competitive intrusion into the Ford circle of relatives, leaving behind physical evidence of its go to.

Skeptics argue that those sightings can be misinterpretations of black bears close by to the vicinity. Regardless of the beginning location, Fouke embraces its legend. Establishments like Peavy's Monster Mart have commercialized the lore, offering an array of products or maybe picture opportunities with the legendary beast's illustration.

In the quit, whether or not myth or reality, the Fouke Monster stays an extended-lasting symbol of Arkansas' rich tapestry of folklore and legends.

California: Dark Watchers

For over 3 centuries, internet site online web page site visitors and citizens close to the Santa Lucia Mountains of California have stated encountering towering, enigmatic figures shrouded inside the night time's shadows. These mysterious entities, called the Dark Watchers, are regularly seen inside the hours flanking dusk and sunrise, gazing from the mountaintops as despite the fact that looking on the arena underneath.

The Santa Lucia Mountains, which run from Monterey County to critical San Luis Obispo County, posed a powerful barrier to Spanish explorers in California's early days. As those explorers navigated the steep terrains within the 1700s, many claimed to come to be aware about the towering shadowy figures and dubbed them "Los Vigilantes Oscuros," translating to "the dark watchers." Centuries later, American settlers traversing these mountains cautioned a similar unsettling feeling of being positioned from the heights.

Descriptions of the Dark Watchers often paint them as tall human-like silhouettes, everywhere from 7 to 15 toes in pinnacle. They're commonly clad in darkish flowing cloaks, don sizeable-brimmed hats, and are every so often seen preserving staves or sticks. Their faces, if any, remain hidden in shadow. Their presence is fleeting, for they vanish if one tries to technique, giving them a trustworthy greater elusive and mysterious nature.

Renowned writer John Steinbeck alluded to the ones figures in his brief story

"Flight." His very very very own mom, Olive Hamilton, emerge as a enterprise organisation believer in their existence. She regularly shared reminiscences of her encounters with those beings in the direction of her treks as a extra younger teacher inside the mountains. According to Thomas Steinbeck, her grandson, she no longer fine discovered them but even exchanged gives with them inside the area of Mule Deer Canyon.

Brian Dunning, writer of Skeptoid Magazine, acknowledges that the genuine nature of the Dark Watchers stays a conundrum. Many feature their sightings to pareidolia, wherein our brain discerns familiar shapes in ambiguous stimuli. Some hypothesize that the sightings are simply optical illusions known as the "Brocken Specter." This phenomenon, recognized in Germany's Harz Mountains, results from the sun casting a magnified shadow of an observer onto the clouds, growing the illusion of a huge, looming determine.

Further theories advise that hallucinations, stemming from exhaustion and reduced oxygen levels at higher altitudes, might be chargeable for the sightings. Another captivating clarification includes infrasound — subaudible frequencies that might invoke feelings of unease or perhaps terror. Experiments have shown that exposure to certain low-frequency sounds can lead individuals to enjoy anxiety, dizziness, and a enjoy of presence.

The real nature of the Dark Watchers remains one of California's most captivating mysteries. Whether they may be products of natural phenomena, mental effects, or a few element absolutely unexplained, their recollections keep to captivate and mystify those who pay attention of them. What's effective is that their legend, exceeded down thru generations, stays deeply rooted inside the cultural material of the area.

Colorado: Slide-Rock Bolter

Nestled within the expansive terrains of Colorado is a legend that has both intrigued and apprehensive its inhabitants for years. The Slide-Rock Bolter, on occasion definitely referred to as the Slide-Bolter, is a mysterious cryptid whose testimonies date returned to the 19th and twentieth centuries. Described as a big creature, reminiscent of a finless dolphin, this beast possesses small eyes, a cavernous mouth, and hand-like hooks at the give up of its tail flipper. Its sheer length is frequently in evaluation to a whole hillside, making it an outstanding sight to count on.

What makes this creature a topic of nightmares is its modus operandi. The

Slide-Rock Bolter is concept to latch onto steep mountain slopes the usage of the hooks on its tail. When the timing is proper, it releases itself, sliding down the slope at excellent speeds to capture and eat any unfortunate beings in its route. With one short waft, it is able to swallow entire companies of humans. After its ferocious attack, the creature searches for a few different hillside to realize to, making ready itself for its next ambush.

According to close with the aid of folklore, the Slide-Rock Bolter became this sort of risk that a park ranger within the San Juan Mountains devised a plan to place an quit to its reign of terror. The ranger located a decoy tourist filled with explosives in a valley beneath the Lizard Head height, a extremely good summit. As anticipated, the Slide-Rock Bolter lunged on the bait, ensuing in a incredible explosion upon consumption. The aftermath of this explosion became so terrific that it leveled a large part of the encompassing region, collectively with 1/2 of the city of Rico. It's believed that this occasion marked the prevent of this unique Slide-Bolter. However, the actual amount of those creatures, inside the event that they truely exist, stays a mystery.

Despite the chilling memories, the authenticity of the Slide-Rock Bolter's existence is widely debated. Many argue that the legend became truly an try to offer an cause of herbal activities alongside facet landslides and avalanches, which had been possibly misunderstood with the aid of the early settlers of the Rocky Mountains. These avalanches and their terrible aftermath need to with out trouble be personified into the testimonies of a large creature that slides down mountains. To this modern, there isn't concrete proof proving the existence of the Slide-Rock Bolter. Yet, its legend stays, serving as a chilling reminder of the mysteries that the Colorado mountains may additionally moreover hold.

Connecticut: Melon Heads

In the folklore of the contemporary era, few creatures are as weird and mystifying due to the fact the Melon Heads. These humanoid entities, characterised by their disproportionately large, veiny heads and diminutive our our bodies, are rumored to lurk inside the secluded geographical regions of the woods, geared up to attack anyone who dares project too close to.

Connecticut, alongside Ohio and Michigan, is a first-rate hotspot for stories of these cryptic beings. Each country has its personal amazing model of the Melon Head legend, with the overall consensus being that they emerged from both an insane asylum or an orphanage following a violent rebellion or catastrophic fireplace. This calamity resulted within the deaths of maximum populace and body of workers, the usage of the surviving Melon Heads to are attempting to find secure haven inside the forests.

Southwestern Connecticut is specifically rife with recollections of those beings. Heavily wooded u . S . Roads inclusive of Zion Hill Road in Milford and Saw Mill City Road in Shelton are believed to be their number one domains.

Appearing as small humanoids with heads that appear to weigh down their our our bodies, Melon Heads are elusive, rarely revealing themselves. They are perception to keep themselves on a diet that consists of small animals, stray cats, and, most disconcertingly, human flesh— predominantly that of young adults. These Melon Heads, consequently, function handy scapegoats for the mysterious

24

disappearances of runaway teens or hikers.

Digging deeper into the annals of statistics, one reveals that testimonies of deformed, reclusive u . S . A . Oldsters have circulated for over a century, reaching as a long way back as nineteenth-century Europe. For instance, a circle of relatives of Melon Heads, called "weeble heads," emerge as rumored to inhabit the outskirts of Risbury, England, spherical 1900.

In america, memories of Melon Heads began to proliferate in Connecticut put up-World War II, a time marked via manner of town to suburban migration. These reminiscences in all likelihood mirror the prejudices and fears that New Yorkers harbored towards the insular rural folks.

Multiple theories try and elucidate the origins of Connecticut's Melon Heads.

Some posit that they're the descendants of a family banished for witchcraft, who resorted to inbreeding for survival, in the long run mutating into Melon Heads. Others propose that they're escapees from Fairfield Hills Hospital, a now-defunct intellectual company, or Garner Correctional Institute, which homes inmates with highbrow health troubles. A variant of this concept proposes that the Melon Heads are the progeny of inmates from an unnamed intellectual organisation that modified into engulfed in flames inside the Nineteen Sixties. Their turn to cannibalism, as a way of survival, is notion to have resulted of their in reality swollen heads.

Similar legends of Melon Heads also pervade the lore of Ohio and Michigan. In Ohio, an evil Dr. Crow is said to have accomplished ghastly experiments on orphans in Kirtland, in the end important to their break out, the orphanage's destruction via hearth, and their subsequent retreat into the barren place. Meanwhile, Michigan's Melon Heads are concept to be sufferers of abuse in an insane asylum in Ottawa County who had been later let loose into the woods.

An interesting legend from the 1980s tells of a hard and fast of Fairfield's Notre Dame High School girls, who encountered the Melon Heads in Trumbull. The girls reportedly came face-to-face with these beings after venturing into the woods, with the Melon Heads eventually the use of off in the women' blue Granada.

Despite the wealth of narratives, the life of Melon Heads remains shrouded in thriller, their tales serving as an eerie testomony to the enigmatic forces that would stay in the shadows of Connecticut's forests.

Chapter 3: Selbyville Swamp Monster

Swamps are regularly steeped in mystery, and Delaware's Selbyville Swamp is not any exception. Shrouded in mists and shadows, those wetlands conceal secrets and techniques and strategies and memories that have been handed down through generations. Local residents whisper of the Selbyville Swamp Monster, from time to time called the Burnt Swamp Monster. Tales of this cryptid have permeated southern Delaware folklore, some attaining back to the Nineteen Thirties.

In one such tale, raccoon hunters embarked on a looking experience into the dense swamp, their hearts whole of choice and exhilaration for a fruitful hunt. Their excursion took a terrifying turn after they had been startled by using using manner of a chilling scream, in comparison to any recounted creature that inhabited the swamp. Rather than confronting the unknown entity that produced the scream, they selected discretion over valor and unexpectedly retreated, leaving the swamp and its secrets and techniques and strategies in the back of.

The tale of their harrowing revel in spread through the network, sending shivers down the spines of people who heard it. Soon, subsequent memories surfaced, attributing mysterious occurrences within the region to this enigmatic creature. Reports ranged from vanishing pets to hard blood lines within the swamp or maybe mutilated cattle. These testimonies brought gas to the fire, captivating the imaginations of the locals and horrifying fear of their hearts.

Eyewitnesses who claimed to have seen the monster supplied diverse descriptions. Some portrayed the monster as a ghostly apparition, a spectral figure that glided thru the swamp, reputedly unaffected by way of way of way of the natural limitations that could avoid a corporeal being. Others painted a extremely unique photograph of a bipedal half of-human, 1/2-beast blanketed in fur, a creature of the earth that prowled the swamp on legs, its intentions unknown.

In a twist to the tale, in 1964, the "Delmarva News" featured a story at the Selbyville Swamp Monster, complete with a photograph that purportedly captured the creature in its herbal habitat. However, this narrative and its accompanying picture have been later uncovered as a concocted story designed to decorate newspaper profits. This revelation robust a shadow of doubt over the life of the swamp monster, however it didn't quash the network's fascination with the creature. Nor did it supply an cause of the bills of sightings and research that predated the newspaper hoax.

Today, questions linger. What had been residents encountering in advance than the 1964 newspaper hoax? And put up-1964, did sightings without a doubt stop, or did the swamp monster surely slip lower returned into the shadows, hidden from prying eyes? Regardless of the anomaly surrounding its life, the Selbyville Swamp Monster stays an important part of Delaware's cryptid lore, a story this is nevertheless informed spherical campfires and in hushed voices within the dead of night time time. The swamp holds its secrets and techniques and techniques and techniques close to, however for the ones brave sufficient to mission into its depths, possibly the reality awaits.

Florida: Skunk Ape

The mysteries of Florida's Everglades have worried locals and tourists alike for generations. Perhaps one of the maximum exciting and elusive mysteries of this massive swampy expanse is the legendary Skunk Ape. This cryptid, much like Bigfoot's Southern cousin, is defined as a massive, upright taking walks ape standing over seven feet tall, protected in hair, and exuding a smelly fragrance reminiscent of rotten eggs, spoiled meals, or decay.

The tale of the Skunk Ape has been passed down through community folklore and gained prominence for the duration of a wave of sightings in the 1960s. These payments came from numerous individuals who ventured into the thick swamps and said encounters with this mysterious beast. Some instructed tales of glimpsing a big, bushy determine in the distance, on the identical time as others stated locating uncommon footprints or smelling its foul stench.

One of the most compelling pieces of evidence for the Skunk Ape's existence is the case known as the "Myakka Skunk Ape" incident in 2000. The Sarasota County Sheriff's Department acquired pix and an nameless letter from an elderly female who claimed to have captured pictures of the Skunk Ape lurking in her out of doors. She stated that the creature were pilfering apples she disregarded. This case has been hotly debated by means of cryptozoologists and skeptics alike, with some offering severa motives for the snap shots, on the same time as others insist on their authenticity.

Beyond those contemporary payments, the Skunk Ape has a deep-rooted records in the folklore of Florida and the encircling Southern states. Tales of a huge, foul-smelling ape-like creature have been advocated inside the area for the cause that instances of the early European settlers. Some argue that the Seminole and Miccosukee tribes have prolonged informed memories of a comparable creature known as Esti Capcaki, translating to "Furry Tall Man" or "Hairy Giant."

These early evaluations, mixed with the multitude of sightings inside the twentieth century, have contributed to the Skunk Ape's legendary reputation in Southern folklore. The creature is stated to roam the swamps and forests, leaving in the back of simplest fleeting glimpses and a smelly odor as proof of its existence.

Despite its prominence in nearby life-style, mainstream technological facts remains skeptical of the Skunk Ape's existence, dismissing most evidence as hoaxes, misidentifications, or folklore. Some suggest that the sightings is probably attributed to American black bears, a number of which might also suffer from mange, giving them an unusual appearance.

Regardless of the controversy surrounding its lifestyles, the Skunk Ape remains an important part of Florida's rich tapestry of folklore and mystery. It continues to capture the imaginations of locals, travelers, and cryptozoologists alike, all of whom assignment into the depths of the Everglades in hopes of uncovering the reality approximately this elusive creature. As the swamp mists upward push and the shadows lengthen, the legend of the Skunk Ape lives on, a protracted-lasting enigma that continues the thriller of the Florida Everglades alive and properly.

Georgia: Altamaha-ha

In the brackish waters of the Altamaha River near Darien, Georgia, locals speak of a mysterious and elusive creature known as the Altamaha-ha. Often affectionately known as "Altie," this river monster is described as being approximately 20-30 ft in length, with a seal-like snout, great flippers, and a sinuous, snake-like body. Its skin is generally endorsed to be a vibrant inexperienced, contrasted thru a mild, yellowish underside.

The legend of the Altamaha-ha has roots relationship once more to the 1830s. Captain Delano is regularly credited with the primary said sighting of the creature, describing a serpentine beast within the river. Since then, numerous sightings have been documented. In the Twenties, timbermen working alongside the river claimed to have witnessed the monster, and their descriptions cautiously resembled that of Altie. A Boy Scout troop referred to a sighting inside the Forties, and inside the Nineteen Fifties, officers from Reidsville State Prison claimed to have visible the creature as well.

The legend of Altie isn't always restrained to historic debts. In 2002, a person boating on the river said a near stumble upon with a 20-foot-prolonged creature that matched the define of the Altamaha-ha.

The Altamaha River itself is a massive body of water, stretching 137 miles and fed thru the Ocmulgee and Oconee Rivers, sooner or later flowing into the Atlantic Ocean. The river has lengthy served as a essential resource for transportation, fishing, and attempting to find the inhabitants of Georgia.

Adding to the appeal of the legend, in 2009, sculpture artist Rick Spears created a lifestyles-sized rendition of Altie based totally mostly on eyewitness debts. This sculpture is proudly displayed within the foyer of the Darien Welcome Center, serving as a testament to the lasting effect this close by legend has had on the network. Visitors are invited to take a selfie with Altie, immortalizing their come upon with this mythical creature of Georgian folklore.

Hawaii: Menehune

Stories of the mysterious Menehune, a race of dwarf human beings, have been woven into the tapestry of Hawaiian mythology and folklore, with each historic and current money owed talking in their life.

The Menehune are described as muscular and brief, with dark or purple pores and pores and skin, long eyebrows, huge eyes, and sticking out foreheads. They make their houses inside the most some distance off regions of the Hawaiian Islands, building shelters out of banana leaves, hole logs, lava tubes, and caves.

Their diet plan consists of fish, ferns, yams, bananas, breadfruit, and taro. The Menehune are probable most well-known for their architectural prowess, having constructed large systems, houses, irrigation structures, stoneworks, and fish ponds—all inside the secrecy of the night time time time.

But in which did the Menehune skip? Ancient folklore indicates that they left their hidden caves and dense forests seeking out a new home, a long way some distance from human civilization. The query of their life stays a mystery to nowadays, and no longer using a concrete evidence to verify or deny their presence in current Hawaii.

In the rich tapestry of Hawaiian folklore and mythology, the Menehune maintain a place of prominence, reputation along exclusive mythical creatures like Sasquatch, El Chupacabra, and the Loch

Ness Monster. These creatures have all been extensive by way of curious minds, regardless of the fact that none have been definitively examined to exist past blurry pix or elusive video photos.

Historically, the Menehune had been believed to have arrived in Hawaii in advance than the Polynesian settlers. They have been professional craftsmen and developers, acknowledged to have constructed temples (heiau), fishponds, roads, canoes, and homes. Some systems that also exist nowadays in Hawaii are said to be the work of the Menehune, along with the Kikiaola or Menehune Ditch on Kauai, and the Alekoko Fishpond, also on Kauai. These structures are a testament to the Menehune's craftsmanship and stand as a photo of their legacy.

In 1820, a census of Kaua'i achieved with the useful resource of Kaumuali'i, the ruling ali'i aimoku (local ruler) of the island, indexed sixty 5 human beings as Menehune. This charming piece of facts raises the query of whether or no longer the Menehune were a real race of people or clearly a manufactured from mythology.

Despite the thriller surrounding their life, the legend of the Menehune remains an crucial part of Hawaiian way of existence, and their legacy lives on within the structures they supposedly left in the back of and the tales exceeded down thru generations. As you explore the expensive landscapes and wealthy records of Hawaii, you may find out yourself questioning if the Menehune are nevertheless obtainable, hidden away in the deep forests and valleys, persevering with to craft their architectural masterpieces beneath the quilt of darkness.

Idaho: Sharlie

The legend of Sharlie, the resident lake monster of McCall, Idaho, has captivated the imaginations of each locals and tourists alike for over a century. Nestled within the majestic landscape of applicable Idaho, McCall is home to the adorable Payette Lake, a glacier-carved body of water that spans over five,000 acres and reaches depths of up to 392 toes in its darkest, maximum secluded elements near the Northwest shore. The lake, named after the French Canadian fur trapper Francois Payette, is surrounded thru towering pines and tranquil Douglas fir wood, creating a picturesque putting for this legendary cryptid.

Local Native American tribes as quickly as knowledgeable reminiscences of an evil spirit that dwelled within the apparently bottomless waters of Payette Lake, caution of its hidden depths and the mysterious creature that lurked under.

The first documented sighting of Sharlie came about in 1920 at the same time as a railroad logging crew positioned what they to start with idea to be a massive log floating inside the frigid waters. To their astonishment, the log commenced to transport in advance, undulating in the water and developing its very personal wake because it unexpectedly disappeared from view.

The sightings of Sharlie endured within the route of the years, with one specially dramatic account taking location in 1944 near the Narrows. A employer of human beings mentioned seeing a creature that modified into at least 35 feet lengthy, with a dinosaur-like head, stated jaw, camel-like humps, and hard, shell-like pores and skin. This account garnered country wide interest, with serpent hunters flocking to the location in hopes of taking snap shots a glimpse of the elusive creature. An article within the nationally allotted Times Magazine in August of 1944 cautioned that 30 human beings had seen the periscope-original head of the creature, which modified into referred to as "Slimy Slim" at the time.

By 1954, the legend of Slimy Slim had firmly taken root inside the community, and citizens felt it become time for his or her community monster to have a more appropriate name. A national contest have come to be held, with the winning get entry to submitted thru Leisle Hennefer Tury of Springfield, Virginia. She cautioned the selection "Sharlie," a reference to a well-known one-liner utilized by Jack Pearl in a radio show. The call stuck, and Sharlie became the cherished unofficial image of McCall.

Over the years, ordinary sightings of Sharlie were reported, with descriptions starting from a 30 to forty-foot lengthy creature with a said jaw, camel-like humps, and hard shell-like skin. Despite the lack of concrete proof, the residents of McCall are steadfast in their belief of Sharlie's life. Her legend keeps to develop, together with a hint of thriller and imagination to the region's folklore and serving as a testomony to the rich information and colourful subculture of McCall, Idaho.

Illinois: The Enfield Horror

The Enfield Horror, a mysterious creature alleged to had been visible with the aid of manner of numerous residents of Enfield, Illinois, within the Seventies, has harassed and concerned locals and researchers alike. According to numerous eyewitness bills, the Cryptid stands approximately 4 feet tall and has an unusual appearance, providing arms connected to its front, 3 legs, and sizable pink eyes, with a frame such as that of a monkey. Those who have encountered the creature describe it as being remarkably agile and emitting screeches harking back to a wildcat.

The first documented encounter with this weird entity changed into made by means of way of Henry McDaniel within the early Seventies. McDaniel described the creature in splendid detail, noting its three-legged stance, brief frame, brief hands, and large crimson eyes. He said how the creature made a hissing sound much like that of a wildcat after he fired pics at it, subsequently fleeing inside the course of a close-by railway embankment with brilliant pace and agility.

Following McDaniel's account, severa one-of-a-kind citizens got here in advance with their own sightings of the Enfield Horror. Among them modified into 10-three hundred and sixty five days-vintage Greg Garret, who claimed that the creature had stepped on his toes, ensuing in his tennis shoes being torn to shreds. However, it's miles nicely surely well worth noting that Garret later admitted to fabricating his tale.

Further research into the Enfield Horror led to the discovery of a series of comparable sightings that occurred inside the nearby village of Mt. Vernon among 1941 and 1942. Locals on the time noted the creature in query due to the fact the Mt. Vernon Monster, and regardless of discrepancies in the descriptions of the 2 entities, some accept as true with they will be one and the same.

The media performed a enormous role in propagating the legend of the Enfield Horror, with numerous news shops covering the regular events. Some speculated that the creature could have been a wild ape or an escaped kangaroo, the latter concept being in particular captivating for the reason that McDaniel, who had as quick as owned a kangaroo as a pup at the equal time as serving inside the army in Australia, modified into adamant that the creature he encountered become now not a kangaroo.

The incident became later used as a case have a observe in a paper on social contagion in 1978, with researchers at Western Illinois University studying the episode as an example of collective conduct and the unfold of "organization feelings" which include panic and hysteria.

While the real nature of the Enfield Horror remains a thriller, its legacy is still a subject of interest and talk amongst each locals and Cryptid lovers to in the intervening time. Whether the sightings had been the cease cease end result of proper encounters with an unknown creature, or genuinely a sequence of fabrications and misidentifications, the Enfield Horror has in reality left an indelible mark at the folklore of Southern Illinois.

Indiana: Beast of Busco

The Beast of Busco, additionally affectionately called Oscar, has been a legendary fixture within the Churubusco community, a small metropolis in Indiana, for many years. This cryptid is stated to inhabit the waters of Fulk Lake, a small frame of water named after Oscar Fulk, the farmer who first cited a sighting of the huge turtle in 1898.

Oscar is not your everyday turtle; he is defined as a large creature weighing around 500 kilos, with a shell that would rival the scale of a automobile rooftop or a massive consuming table. Despite Fulk's attempts to capture the creature, it remained elusive, disappearing from sight and leaving at the back of a path of thriller and nearby lore.

The tale of the Beast of Busco took a massive flip in 1948, at the same time as two guys fishing in Fulk Lake claimed to have visible a large snapping turtle, which they identified as Oscar. Soon after, one in every of a kind men repairing a roof additionally stated a sighting of the beast. The description of Oscar as a turtle with a shell longer and bigger than a mean rowboat, a head the dimensions of a child's, and a weight starting from 4 hundred to 500 kilos in addition solidified its presence inside the nearby folklore.

The lake's owner at the time, farmer Gale Harris, became so captivated through the tales of Oscar that he released right into a mission to seize the beast. The story of his quest spread a long way and huge, drawing crowds and media interest to Churubusco. Unfortunately, regardless of the huge efforts located into the hunt, along with the use of traps, hooks, or perhaps attempts to drain the lake, Oscar remained elusive.

Today, extra than half a century later, Churubusco despite the fact that proudly consists of the pick out of "Turtle Town U.S.A." The city will pay homage to its mythical inhabitant with an annual occasion called Turtle Days. This 4-day celebration abilties turtle races, turtle sculpture judging, and a variety of carnival attractions, drawing people from all through to take part within the festivities.

The thriller of Oscar's whereabouts remains a topic of speculation and intrigue. While a few receive as real with that he has left the lake, others are satisfied that he still is living inside the depths of Fulk Lake, waiting to make an look. Regardless of in which Oscar can be, one factor is for positive – his legend continues to thrive, including a completely

particular and interesting bankruptcy to
the folklore of Churubusco, Indiana.

Chapter 4: Van Meter Monster

In the quiet, humble town of Van Meter, Iowa, a legendary creature called the Van Meter Monster has been the supply of every terror and fascination because of the reality its first recommended sighting in 1903. Described as an 8-foot tall humanoid with bat-like wings and a horn on its head that emits a stunning white moderate, this cryptid has stressed and captivated the imaginations of locals and cryptozoologists alike for over a century.

The first splendid sighting of the Van Meter Monster changed into made with the aid of U.G. Griffith, a prominent businessman inside the area, who became baffled to look a half-human, half of-winged creature leaping from constructing to building one night time time. Dr. Alcott, any other first rate determine within the network, moreover encountered the monster at night time and went up to now as to fireside at it multiple times, but to no avail. The creature seemed impervious to human intervention, as confirmed thru any other incident regarding Clarence Dunn, a monetary organization manager who confronted the beast and fired at it together with his shotgun, handiest to discover huge footprints outdoor the subsequent morning.

Among the witnesses have turn out to be O.V. White, who observed the Van Meter Monster perched atop a mobile phone pole. Despite his tries to harm the creature, it retaliated by way of the usage of freeing an lousy scent in his direction. Sydney Gregg, White's neighbor, witnessed the monster climb down from the pole and fly thru the city towards the vintage coal mine, that could later emerge as a amazing area in the lore of the Van Meter Monster.

In the following days, the city have become abuzz with reports of "flying subjects" in the location, raising questions on whether or not or now not the Van Meter Monster had once more, or if the sightings may be attributed to particular large birds or even supernatural entities. The mystery surrounding this cryptid has continued, prompting authors Chad Lewis, Noah Voss, and Kevin Lee Nelson, with assist from close by librarian Jolena Walker, to delve deeper into the legend of their ebook "The Van Meter Visitor."

Through considerable studies and interviews with close by residents, the authors exposed stories handed down through generations about the monster and its haunts. The historical mine wherein the creature have become closing seen changed into additionally explored, providing a tangible link to the activities of 1903. Though the authors observed no concrete evidence to verify the lifestyles of the Van Meter Monster or debunk it as a hoax, the fact stays that the story has withstood the check of time, turning into an essential part of the metropolis's facts and lifestyle.

The Van Meter Monster, much like excellent cryptids, continues to be a supply of intrigue and hypothesis, with the questions surrounding its beginning, lifestyles, and disappearance final unanswered. While some may furthermore brush aside it as a figment of imagination or an complex hoax, others view it as a fascinating glimpse into the world of the unknown, wherein some thing is possible, and the street among reality and mythology is blurred. Regardless of wherein one stands on the problem, the legend of the Van Meter Monster stays a protracted-lasting and charming story that has cemented its location within the folklore of Iowa.

Kansas: Sink Hole Sam

The legend of Sink Hole Sam is one which has captured the imaginations of locals and outsiders alike. The monster, in keeping with folklore, is stated to live in an underground cavern associated with Inman Lake, specifically part of the lake referred to as "the big sinkhole." It is in this mysterious and possibly ancient cavern that Sam has allegedly made its home.

The first said sighting of Sink Hole Sam turn out to be by using men fishing on the massive sinkhole. Their tale prompt a sequence of sightings, consisting of one through George Regehr and Albert Neufeld. These guys claimed that that that they had encountered a creature approximately 15 toes in length, with a diameter similar to an car tire. Neufeld even stated having taken pics on the creature, although it seemingly escaped unhurt.

In the years following the ones initial sightings, reviews of monster sightings advanced. Many of these sightings had been made thru folks that had traveled to the area particularly within the hopes of catching a glimpse of the notorious beast. Despite the diverse money owed of encounters with Sam, no bodily evidence has been produced to verify the creature's life.

The origins of the legend are as an opportunity unsure. A possible cause for the tale's genesis might be related to a spoof through a Kansas newspaper columnist named Ernest Alva Dewey. In a piece of writing published inside the Salina Journal Sunday, Dewey claimed that he and a Dr. Erasmus P. Quattlebaum (a fictitious individual) had investigated the legend and decided that the creature became a "foopengerkle," a meant species neighborhood to underground Kansas caverns. According to Dewey, the Big Sinkhole became in reality Sam's "above-floor swimming pool."

The article was a satire, however it however introduced hobby to the legend and drew curious people to Inman Lake. There, they was hoping to seize a glimpse of the foopengerkle, which had with the useful resource of then been dubbed Sink Hole Sam. Notably, Mil Penner, a community farmer whose family had lived inside the location for the purpose that 1874, wrote of 1 such occasion in his e-book "Section 27: A Century on a Family Farm." He described dozens of vehicles parked at the edge of the Big Sinkhole, their occupants hoping to become aware of the creature.

Interestingly, but the capability origins of the legend in a satirical article, many "responsible residents" of the area hold that Sink Hole Sam is more than best a humorous tale. For instance, Mary Kay Flynn, a reporter for the Newspaper Enterprise Association, wrote of numerous locals who defined the creature in first rate element, with a few even likening it to Kansas' model of the Loch Ness Monster.

Regardless of its origins, the legend of Sink Hole Sam continues to captivate people to in the meanwhile. Some speculate that the creature is probably a large snake, together with a boa constrictor or python, launched into the wild by way of manner of an uncommon pup owner. Others maintain that Sam is a completely unique creature, akin to the Loch Ness Monster. The truth stays elusive, but one element is for high best: the legend of Sink Hole Sam can be very plenty alive in the minds of people who recognize it.

Kentucky: The Kelly Little Green Men/Hopkinsville Goblins

On August twenty first, 1955, the arena became brought to "little green guys" via the well-known Kelly-Hopkinsville incident. To a few, this UFO incident isn't whatever extra than an overreaction fueled with the resource of moonshine to an animal attack; others undergo in mind it compelling proof of alien contact.

The Bizarre Encounter with Hopkinsville Goblins

On August twenty first, 1955, Billy Ray Taylor have come to be journeying a chum named Elmer "Lucky" Sutton in the small town of Kelly. As he went out of doors to the well to collect a few water, Billy noticed a few detail flying for the duration of the sky. He later described that item as "real shiny" with rainbow-colored exhaust. Panicked, Billy fled indoors and told his spouse and the Sutton circle of relatives that he had actually visible a UFO. At first, nobody took him significantly, however that changed while the dogs started barking, and some thing, or someone, become coming near the residence.

The eleven witnesses later described to police what they observed in terrifying, vibrant terms. The invaders had lengthy arms with talons and round "oversized" heads. Everything about them appeared to glow and shimmer in the darkness. Their our bodies sparkled as they had been fabricated from "silver steel," and their eyes had "yellowish light."

Billy and Sutton grabbed guns and started capturing on the cryptids. Aliens drew nearer over the subsequent few hours, but they in the end retreated. They flipped up into the trees to break out. One reached down and grabbed Billy's hair. Finally, when the stumble upon became over, numerous people fled into city in a vehicle to invite the police for help.

The police and their leader, Russel Greenwell, arrived on the Sutton farmhouse to investigate, however they did now not find clean evidence of "little silver guys."

Theories About the Incident

The validity of this entire stumble upon become confused right now. Neighbors idea of the whole situation as a drunken debacle, and a few people doubted the honesty of the witnesses. And even as the Sutton family started charging admission to their farmhouse to earnings off interest inside the story, any very last goodwill in the course of them disappeared.

There are more than one theories approximately the Kelly-Hopkinsville incident. One concept is that the humans at Sutton farmhouse mistook a Great Horned Owl for an alien in the darkness. Great Horned Owls have spherical heads, glowing eyes, and lengthy wings. These trends kind of in shape the outline of the so-referred to as Hopkinsville Goblins.

Another precept furthermore consists of an animal. A sheriff from a close-by metropolis, Arthur "Hoss" Cansler, joined the network police for the research at the Sutton farmhouse. According to his document, absolutely everyone became drunk, and a few tossed a cat onto a display display to frighten the human beings interior. Again, this concept is based mostly on all of the witnesses being inebriated; Joann Smithey, who arranges the every year "Little Green Men" Days Festival, believes this idea invalid.

Several a few years have exceeded because the Kelly-Hopkinsville come across, and these days, the reality about what happened is extra elusive than ever. However, one trouble is exceptional; this incident helped solidify the idea of aliens as "little green guys." This is because of the reality newspapers used the time period "little inexperienced men" at the same time as telling the story, however the fact that the Suttons and Billy defined the creatures as "little silver men."

Louisiana: Rougarou/ Loup-Garou

In the colourful, mystique-stuffed lands of Louisiana, locals whisper about an ominous cryptid referred to as the Rougarou, a creature steeped in Cajun folklore and notorious for its fearsome look and searching prowess. This 1/2-wolf, half of of of-guy creature allegedly prowls the swamps, fields, and outskirts of old fashioned Louisiana cities, continuously looking for its subsequent unlucky prey.

The name "Rougarou" is derived from the French term "loup-garou," with "loup" meaning wolf, and "garou" translating to "guy who transforms into an animal." This fearsome creature is regularly described as a humanoid decide sporting the top of a wolf or canine, entire with glowing red eyes that pierce the darkness of the Louisiana night time. A remarkable difference a number of the Rougarou and the imperative werewolf is that the Rougarou does now not require the presence of a entire moon to convert. It keeps its humanoid form during sunlight hours and only morphs into its terrifying, wolf-like visage beneath the cloak of night time.

Folkloric narratives surrounding the Rougarou are as rich and sundry because the manner of lifestyles from which they emerged. One famous version of the story indicates that the creature specially preys

66

upon Catholics, particularly folks who defy the guidelines of Lent. According to this rendition, the Rougarou will hunt down and consume the blood of non-compliant Catholics inside the path of the Lenten season. Another captivating generation of the story shows that the curse of the Rougarou lasts for a precisely counted one zero one days. In this narrative, as speedy as the creature has efficaciously drawn blood from its sufferer, the curse is transferred to the unfortunate soul, simultaneously relieving the actual bearer of the curse.

An thrilling defensive degree accompanied with the resource of the locals to protect themselves from the menacing Rougarou includes putting thirteen small gadgets, which include pebbles or pennies, at one's doorstep. Folklore has it that the Rougarou is incapable of counting past the huge variety 12. Consequently, even as it techniques a residing to say its subsequent sufferer, it turns into sincerely burdened and fed on with the resource of the undertaking of counting the devices, subsequently finding itself trapped in an endless cycle of counting until the primary mild of sunrise forces it to retreat back to the swamps from whence it came.

The legend of the Rougarou is so deeply ingrained in Louisiana's cultural tapestry that it's been immortalized in numerous types of famous subculture. This consists of the as quickly as a 12 months Rougarou Festival held in Houma, a testomony to the cryptid's enduring appeal. Furthermore, the Audubon Zoo in New Orleans is domestic to a Rougarou showcase that abilities an intensive statue of the creature, taking pictures its swampy essence in all its glory.

In end, the legend of the Rougarou is a charming tapestry woven from the threads of French folklore and Cajun life-style, developing a tale rich in subculture and thriller. The Rougarou remains an iconic discern in Louisiana's cultural landscape, a testomony to the power of storytelling and the long-lasting appeal of the mysterious and the unknown.

Maine: Specter Moose

In the dense forests of Northern Maine, an implementing and mysterious creature has captivated the imaginations of locals and hunters alike because the past due nineteenth century. Known because the Specter Moose, this extraordinary animal has been the problem of severa sightings and folklore, weaving itself into the very fabric of network legend.

The first recommended sighting of this huge moose came from Joe Francis and his brother Charlie in October 1891. Described as a first rate and nearly otherworldly beast, the Specter Moose stands at an imposing pinnacle of 10 to fifteen ft tall. With its dusty white coat that looks to emit a faint, ghostly glow, the creature is virtually a sight to behold.

However, the Specter Moose is not merely a surprise to observe; it is also endowed with an array of supernatural capabilities. Its heightened experience of scent and supernatural hearing make it an elusive prey for any hunter formidable enough to pursue it. With its massive antlers, the moose is a powerful opponent, towering over any capability threats. Adding to its mystique, the Specter Moose has the capability to disappear and reappear at will, similar to a phantom. This, coupled with its functionality to section via sturdy objects, has left many to question whether or not the creature is of this international.

The Specter Moose has been sighted dozens of instances inside the direction of

Maine considering the truth that its first look in 1891. Each come across with this awe-inspiring creature has most effective brought to its legend, with the ones fortunate enough to have witnessed it claiming that no particular moose compares in weight, stature, or the grand unfold of its antlers.

But what makes the Specter Moose definitely precise is the manner it has captured the creativeness and hobby of individuals who pay hobby its tale. The memories of this ghostly creature have been surpassed down from technology to era, which includes to the rich tapestry of neighborhood folklore and legend.

Over the years, many have attempted to provide an reason behind away the sightings of the Specter Moose as simply encounters with albino moose. These uncommon animals, even as placing in look with their white coats, lack the sheer length and supernatural capabilities attributed to the Specter Moose. Additionally, even as albino moose typically have purple eyes, the Specter Moose is said to have brown eyes, in addition distinguishing it from its albino opposite numbers.

Others have posited that the Specter Moose might be suffering from a scenario because of an infestation of winter ticks, that could bring about a lightening of the animal's coat. However, this case additionally typically reasons the moose to lose most of its hair, leaving its body emaciated and thin - a far cry from the robust and exceptional creature defined in sightings.

Regardless of the skepticism surrounding its life, the locals of Northern Maine take the legend of the Specter Moose very severely. The creature has turn out to be a image of the location's wealthy information and cultural information, a testomony to the energy of folklore in shaping a community's identification.

In quit, the Specter Moose of Maine is a captivating and mysterious creature that has captured the creativeness of many. Whether it is sincerely a big breed of moose with a unprecedented colour pattern or a paranormal, ghostly animal that has roamed the area for over a century, the legend of the Specter Moose remains a deliver of surprise and intrigue for folks who listen its tale.

Maryland: Goatman

The Goatman of Maryland is a creature shrouded in thriller and steeped in community legend. With its half-guy, half of-goat look, it represents a ugly amalgamation that has haunted the collective creativeness of companies in

Prince George's County, Maryland, in addition to Kentucky and Texas.

The Goatman is thought to lurk within the shadows below bridges, conceal inside the darkness, and stalk parked cars and unsuspecting passersby. It is stated to be liable for a gaggle of violent actions beginning from unexpected dual carriageway injuries to animal mutilation or even belongings destruction. With its huge length, foul fragrance, and terrifying look, the Goatman is a determine that has struck fear into the hearts of locals and intrigued the curious minds of folklore enthusiasts.

Legend has it that the Old Alton Bridge in Texas has been nicknamed "Goatman's Bridge" because of severa sightings of the creature inside the area. Some speculate that the Goatman can be related to the infamous Bigfoot, sharing the same family of mysterious and elusive beasts that roam the rural areas of the US. However, what gadgets the Goatman apart is its weird basis story that hyperlinks it to a scary test lengthy long gone wrong on the Beltsville Research Agricultural Center. According to this version of the legend, the Goatman have become the surrender result of a genetic take a look at that combined human and goat DNA, ensuing in the terrifying creature we realise nowadays.

Despite the numerous sightings and reports, the proper basis and lifestyles of the Goatman stay a topic of discussion. The first media component out of the Goatman dates another time to October 27, 1971, inside the Prince George's County News, wherein the creature end up mentioned alongside special network folklore. This preliminary document laid the idea for what would possibly turn out to be a well known and fear-inducing legend.

Interestingly, the Goatman's tale obtained similarly traction with an editorial published inside the Washington Post on November 30 of the identical year. The article superb an incident regarding a own family's search for their missing domestic dog, Ginger, who changed into later placed decapitated near Fletchtown Road. The connection a number of the Goatman and the home dog's grotesque future delivered gas to the fireplace of the legend, with reviews of extended sightings of the "animal-like creature that walks on its hind legs" within the region of Fletchtown Road.

The Goatman legend has been surpassed down from generation to generation, taking pictures the imagination of locals and visitors alike. Some have even taken to "Goatman searching" in an try and trap a glimpse of the elusive creature. The legend has been a staple of close by folklore for many years, evolving over time and incorporating new elements that add to its mystique.

The numerous bills of the Goatman's look variety, with a few describing it as having a human face but with a frame included in hair, even as others claim it more closely resembles a creature that is both a humanoid blanketed in hair or a human with the bottom half of of a goat.. Regardless of its precise look, the Goatman has emerge as an iconic determine in the global of American folklore, representing the extraordinary and unknown that lurks in the shadows of rural organizations.

Chapter 5: Dover Demon

The Dover Demon remains one of the most mysterious and charming cryptids advised within the city of Dover, Massachusetts. This enigmatic creature became sighted three times over the course of nights in April 1977, sparking a flurry of hobby and research that keeps to this contemporary.

The first stumble upon with the Dover Demon took place at the night time of April 21st, 1977, at spherical 10:30 PM. A youngster named Bill Bartlett modified into using along Farm Street with pals even as their car's headlights illuminated a unusual decide crouching on a damaged stone wall. The creature have become in contrast to some difficulty Bartlett had seen earlier than — it had a "watermelon-common" head, large, gripping palms and feet, spindly, skinny arms and legs, and hairless, orange-tinted difficult pores and pores and skin. Although Bartlett's friends did now not see the creature because it rapid scampered away, his centered description of the entity may want to display vital in identifying the cryptid in a while.

Later that equal night time, in the dark, the creature became sighted once more. This time, the eyewitness emerge as John Baxter, a fifteen-twelve months-vintage teenager who modified into on foot home from his woman pal's residence. As Baxter traversed Miller Hill Road, he found a mysterious parent inside the distance. At first, he believed it to be a pal who lived close by and called out to it. However, the figure all of sudden ran away, prompting Baxter to offer chase. He in the end located the creature crouched on pinnacle of a rock. According to Baxter's account, the creature had feet that regarded to be "molded" to the form of the rock and resembled a monkey with a "discern-8" shaped head.

The very last said sighting of the Dover Demon took place the following night time time time. Will Taintor, 18, modified into using down the road with 15-three hundred and sixty five days-antique Abby Brabham when they got here across the creature on Springdale Avenue. Their description of the weird entity matched the ones given through Bartlett and Baxter. However, Brabham added a unique element – the creature had vivid, glowing green eyes.

The similarities between the debts of the three separate sightings are setting. All three witnesses referred to seeing a creature with a huge head, thin limbs, and no visible hair. Moreover, the truth that the sightings befell indoors a 24-hour period in the equal city provides to the credibility of the witnesses. All 3 encounters have been endorsed to the community police, who accomplished an research into the sightings. As a part of this studies, the witnesses were asked to attract the creature that they had seen. Despite being at a loss for words one by one, each drawing depicted a creature with splendid similarities, with the nice large difference being Brabham's insistence that the creature had green eyes.

The neighborhood newspaper became brief to pick up at the story, coining the call "Dover Demon" to offer an reason behind the mysterious cryptid. Yet, however the attention the case obtained, no concrete evidence of the creature became ever decided. The lack of physical evidence, coupled with the fact that there had been no reported sightings of the Dover Demon at the same time as you bear in mind that 1977, has led some to impeach the validity of the witnesses' payments. However, the suitable descriptions provided through the witnesses, combined with the similarities between their debts, advise that they did definitely see some thing uncommon on those nights in April 1977. Whether the Dover Demon is a however-to-be-positioned species, an alien traveler, or something else actually remains one of the first-class mysteries within the global of cryptozoology.

Michigan: Dogman

The Michigan Dogman, a completely unique and notorious cryptid, has been thrilling Michiganders and beyond for over a century, with its towering presence, wolf-like abilities, and human-like stance. This cryptid is frequently defined as fame upright like someone, with the bushy body of a dog or wolf and a face that is a mixture among canine and ape. Witnesses frequently describe it as which encompass a wolf, but there were money owed of it appearing more like a wolfhound or greyhound, with fur hues beginning from black to silver.

The most terrific sightings of the Dogman date once more to 1961, with different large encounters said inside the mid-1970s, 1987, and 1993. While Michigan holds the bulk of these recollections, comparable testimonies have emerged from unique states like Colorado, Texas, and Wisconsin.

However, the Dogman did now not in fact capture the general public's hobby until a song titled "The Legend" changed into broadcasted by way of a community radio station on April 1st, 1987. Initially supposed as a prank, the station became surprised to gather calls from credible and intense-minded people claiming that they'd encountered the Dogman and have been searching for to hook up with others who had similar reviews.

A harrowing account of a Michigan truck reason pressure named Joe Barger provides each different layer to the enigmatic lore of the Michigan Dogman. Barger, a six-three hundred and sixty five days military veteran, encountered the creature whilst driving via the Manistee National Forest. His tale, shared at the YouTube channel "What Lurks Beneath," has garnered massive hobby, losing slight at the unsettling nature of these encounters.

Barger became transporting a load of paper thru the forest while he stopped to restore an air leak. It have grow to be then that he heard an unnatural vocalization and noticed shadowy figures within the woods. Initially dismissing it as a undergo, he fast determined out that he changed into face to face with a few thing a protracted manner extra sinister.

As he resumed using, Barger discovered a large wolf head beside his window, with the creature preserving tempo collectively along together with his truck. The Dogman, as Barger described, changed into as a minimum ten ft tall with sharp teeth, amazing white fangs, pointed ears, and yellow eyes. In a kingdom of surprise and worry, Barger instinctively reached for his handgun and fired photographs on the creature, which appeared to have delivered it down.

But the tale doesn't cease there. Barger claims that he desired to affirm the creature's dying, however upon returning to the internet site, the Dogman emerge as lengthy beyond, modified by way of using the use of human beings in a jeep who claimed to have seen bears stopping. This led Barger to make investments that there might be extra Dogmen inside the place.

The aftermath of Barger's stumble upon was stressful, with nightmares plaguing him for months. In a bizarre twist, Barger claims that he became detained with the useful aid of federal government, who confiscated his gun and in quick close to off his financial group account, warning him to prevent speaking approximately the incident. Barger alleges that the federal authorities is aware about the Dogman's lifestyles and is maintaining it beneath wraps for capability military use.

Interestingly, the Michigan Dogman is stated to appear in a 10-365 days cycle, with the primary recorded sighting dating once more to 1887, whilst lumberjacks claimed to have seen a creature with someone's body and a dog's head. This cyclical sample, culminating in years finishing in '7', suits with Barger's come upon, which passed off in 2017.

The lore of the Michigan Dogman continues to conform, with every account consisting of every other layer to the mystery. Whether the Dogman is a figment of creativeness, a misidentified animal, or something a ways more mysterious, it maintains to capture the interest and fear of folks who dare to delve into the folklore of Michigan's most notorious cryptid.

Minnesota: Wendigo

In historic North American folklore, a terrifying cryptid known as the Wendigo is said to stand up from darkish magic or cannibalistic practices. This creature is deeply embedded inside the legends of severa Native American tribes, with the maximum regular stories stemming from the Great Lakes Region of Minnesota's northern woods, extending as an awful lot as important Canada. Wendigos are continuously related to acts of cannibalism, murder, and unbridled greed, popularity because the embodiment of these heinous movements.

Appearance

Aesthetically, the Wendigo is a grotesque and repugnant creature, bearing humanoid shape but endowed with animalistic attributes which encompass an elongated tongue, razor-sharp claws, and elongated, yellow fangs. In terms of stature, the Wendigo towers over an average human, however its frame is emaciated and haggard, reflecting its perpetual and insatiable starvation for human flesh. Its eyes, a fiery pink, are recessed deep inside their sockets, their luminosity rendering the creature with out difficulty identifiable within the darkest depths of night time.

Albeit a good deal less frequently stated, folklore frequently portrays Wendigos as being shrouded in a layer of ice.

Myths and Folklore

The stories and myths surrounding the Wendigo are each severa and complicated. Some legends purport that the Wendigo isn't a separate entity from human beings, but instead a massive shape that people can redecorate into as a punishment for their moral transgressions, most significantly cannibalism. Engaging in cannibalistic acts supposedly awakens a voracious urge for food for human flesh, compelling the transformed human to are seeking out others.

Other versions of the legend posit that humans can come to be Wendigos via direct interaction with the creature. In these narratives, the Wendigo has the capacity to infiltrate human goals, in the long run assuming control of the character's body and identity.

Despite variations in man or woman money owed, a commonplace thread that binds the ones memories collectively is the depiction of the Wendigo as a primeval, malevolent beast that is aware of handiest cruelty and savagery. In more present day renditions of the legend, Wendigos are every now and then depicted as aliens, with certain narratives drawing connections most of the creatures and unidentified flying items (UFOs).

Wendigo Psychosis

The phenomenon of Wendigo Psychosis is a concept regarded to numerous agencies dwelling within the Northern Ojibwa area. This state of affairs, described as a highbrow or psychological ache, changed into the mission of numerous clinical research in the end of the Nineteen

Sixties. To the local tribes, Wendigo Psychosis is basically taken into consideration a non secular illness, one that originates from malevolent forces lurking in the wooded region. As the situation progresses, the individual is believed to little by little lose their humanity, in the end metamorphosing right into a Wendigo.

One particular take a look at finished inside the Nineteen Sixties postulated that the onset of this case can be attributed to elements which encompass continual loneliness and famine, with medical practitioners advocating for modern-day medicinal interventions as adverse to traditional indigenous restoration practices.

Historical Accounts

Historical facts file severa encounters and modifications concerning the Wendigo, with a number of the earliest debts being chronicled through Jesuit missionaries within the direction of the seventeenth century. In 1907, Algernon Blackwood, an Algonquin local, penned a quick tale detailing a warfare of phrases with the cryptid.

These records, mainly the ones stemming from the Algonquin human beings, are held in high regard, as are the money owed documented through the usage of the use of missionaries, each of which feature credible tales to the lifestyles of the Wendigo.

However, as is the case with many legends and folklore surrounding cryptids, concrete proof of the Wendigo's lifestyles remains elusive, depending mostly on eyewitness debts and anecdotal proof. Despite this, the chilling and frightening nature of the Wendigo continues to captivate the imaginations of present day-

day-day-day hunters, hikers, and adventure-seekers alike.

Mississippi: Witch of Yazoo

Nestled inside the coronary coronary heart of Mississippi, Yazoo City holds a century-vintage legend that intertwines thriller, vengeance, and hearth. This story, handed down through generations, speaks of an eerie figure called "The Witch of Yazoo" or "The Chain Lady" to 3, and her wrath that allegedly left the city in ashes.

The Yazoo River, winding its way near the town, have turn out to be as soon as home to a weird and reclusive lady. Known to the townsfolk as a witch, she resided in solitude close to the riverbanks, shrouded in a cloak of thriller and malice. Legends referred to her luring unsuspecting fishermen from the river, quality to scenario them to torture and torment.

The mystery surrounding the witch grew at the same time as neighborhood sheriff and his deputies stumbled upon an eerie scene in her residing house at some point of the past due 1800s. Following the claims of Joe Bob Duggett, a boy who claimed to have heard otherworldly moans emanating from her house, the lawmen breached the threshold of the witch's residing. The absence of the witch modified into as palpable because the presence of macabre symbols, skeletons placing from the rafters, and the maddening whirl of 1/2-starved cats.

The witch changed into ultimately discovered sneaking away into the depths of the swamp that hugged the banks of the Yazoo River. The sheriff and his men gave chase, ultimately catching as a good deal as the witch as she decided herself ensnared in the clutches of quicksand. It changed into proper right here, in the moments in advance than she vanished under the swampy abyss, that the witch issued her ominous threat – she vowed to break unfastened from her grave and set Yazoo City ablaze on the morning of May 25, 1904.

The witch's vow modified into ignored due to the fact the rantings of a deranged girl, and life in Yazoo City resumed its ordinary course. But on May 25, 1904, the impossible passed off. A catastrophic fireplace swept via the city, eating the entirety in its direction and reducing Yazoo City to smoldering ruins. The townspeople had been left in marvel and disbelief as they bore witness to the devastation that pondered the witch's vengeful promise.

The subsequent day, a group of metropolis elders made their manner to the Glenwood Cemetery, wherein the witch have been laid to relaxation. Their discovery modified into chilling – the heavy chains that were positioned round her grave as a symbolic gesture to include her malevolence were broken. The as soon as-strong links now lay in disarray, adding a tangible layer of credence to the witch's curse.

Today, the tale of the Witch of Yazoo stays an critical a part of the metropolis's folklore. It is a story that has advanced through the years, growing richer in detail with each retelling. While a few regard the legend as an insignificant myth, others acquire as actual with it serves as a testomony to the mysterious forces that now and again intertwine with our life, leaving in the back of an indelible imprint on statistics and memory. The grave of the witch, encircled thru the use of the now-damaged chains, stands as a silent sentinel in Glenwood Cemetery – a stark reminder of the eerie tale that has grow to be Yazoo City's maximum haunting legacy.

Missouri: Momo

For a long time, Missouri's wooded areas and lonely usa of america roads have echoed with whispered memories of a terrifying creature — Momo, or the "Missouri Monster." The creature's lore has grown over time, with its legend cautiously intertwined with the dominion's statistics.

In the summer time of 1971, a chilling come across came about that might emerge as the bedrock of the Momo myth. Joan Mills and Mary Ryan, on the identical time as the use of along Highway 79 close to Louisiana, counseled witnessing a frightening creature that regarded a annoying mixture of ape and guy, emitting unsettling gurgling noises. Their description painted a photograph of a hulking, furry humanoid, now not too unique from the famed Bigfoot of the Pacific Northwest, however with eerie otherworldly competencies. Momo's description typically consists of sparkling orange eyes, a unique pumpkin-commonplace head, 3-fingered arms, and its unusual three-toed footprints.

The preliminary sighting with the useful resource of Mills and Ryan became exceptional the beginning. The following summer season, in 1972, more youthful Terry and Wally Harrison had an extraordinary greater harrowing revel in. While gambling of their out of doors in Louisiana, the youngsters were startled via a massive, black, furry parent lurking via a tree. The creature, believed to be Momo, stood approximately six or seven ft tall, its face hidden below a thick mane of hair. A specially distressing element come to be the sight of the creature carrying what regarded to be a vain dog, leaving the area people in each terror and sorrow.

Such sightings persevered, with residents of Louisiana or maybe St. Charles County recounting their very non-public eerie encounters. Witnesses often located a lingering stench paying homage to rotting flesh, a trait many function to Momo.

The wave of Momo sightings inside the early Seventies drew big media interest, with journalists and researchers from all over the kingdom descending upon Louisiana. Bigfoot fanatics and investigators sought to locate the reality behind those claims, with many focusing at the creature's super three-toed footprints. Among the researchers have become Hayden Hewes, the director of Sasquatch Investigations of Mid America. Hewes stated the real sincerity and exuberance of the witnesses he interviewed, further deepening the mystery surrounding Momo.

While many researchers sought hyperlinks amongst Momo sightings and UFO interest, the proof have become inconclusive. Yet, the ones investigations did set up a sample of Bigfoot-like creature sightings migrating within the direction of the u . S . A . From the Pacific Northwest to the Southeast.

Over the years, Momo's notoriety has pretty waned, transitioning from a chilling truth to a folklore staple in Missouri. However, for individuals who professional it firsthand and those who continue to be intrigued with the aid of using the inexplicable, the legend of Momo stands as a testomony to the mysteries the world though holds. Whether Momo end up a misunderstood black undergo, a series of hoaxes, or a real unknown entity remains a problem of dialogue and fascination.

Montana: Shunka Warak'in

The Shunka Warak'in is a fascinating and mysterious creature that has captured the creativeness of many in Montana and beyond. Described as a wolf or hyena-like cryptid, witnesses declare it is nearly black in color, with excessive shoulders and a sloped decrease back much like a hyena.

The first recorded sightings of the Shunka Warak'in date decrease returned to the Eighties even as settlers first made their manner to the lower regions of Montana. It modified into ultimately of this time that many Native American tribes inside the area named the creature "Shunka Warak'in," which translates to "wearing off puppies" in reference to its addiction of stealing dogs from camps within the useless of night.

One of the most exceptional bills of the Shunka Warak'in come to be recorded in 1886 inside the Madison Valley of Montana. Israel Ammon Hutchins, a settler within the region, cited that an unknown creature became attacking his farm animals, along thing the ones of various farmers and ranchers close by. Hutchins defined the creature as dark and canine-like, with a scream in assessment to a few issue he had ever heard. After a failed try to shoot the creature, ensuing inside the accidental death of considered one of his cows, Hutchins end up in the end able to kill the cryptid. The carcass turned into then traded to Joseph Sherwood, a businessman and taxidermist, who established the creature and displayed it in his grocery maintain/museum in Henry Lake, Idaho, naming it "ringdocus."

The "ringdocus" become displayed inside the museum until as a minimum the Eighties, and then it mysteriously disappeared. The first-rate bodily evidence that remains of the creature is a black and white photograph published inside the autobiography of Ross Hutchins, the grandson of Israel Ammon Hutchins. In the picture, the creature appears wolf-like, however with a wonderful face shape and arched decrease back that sets it aside from ordinary wolves. Some hypothesize that the creature is probably a Shunka Warak'in, a non-wolf canid creature from Native American folklore, infamous for its penchant for abducting dogs.

Curiously, once the "ringdocus" went missing, Jack Kirby, a descendant of Israel Hutchins, exposed that the collection of taxidermy artifacts from Sherwood's museum had decided a brand new domestic on the Idaho Museum of Natural History in Pocatello. Within this series end up the Shunka Warak'in. Kirby managed to regular a mortgage of the creature for the Madison Valley History Museum located in Ennis, Montana, wherein it is been a featured display off for over ten years.

Chapter 6: Alkali Lake Monster

Alkali Lake, positioned inside the Sandhills of Northwestern Nebraska and additionally known as Walgren Lake, spans a tremendous eighty to one hundred acres. This big body of water has captivated the imaginations of locals and location traffic alike, basically because of its connection with the Alkali Lake Monster, a mysterious creature said to inhabit its depths.

The Alkali Lake Monster is often portrayed as a wonderful beast, measuring spherical 40 ft in period, with physical attributes similar to an alligator, however distinguished via way of its difficult, greyish-brown cowl. A singular and defining feature of this creature is the horn-like form located between its nostrils and eyes, setting it apart from any identified animal species.

The first formal documentation of this enigmatic creature can be traced decrease once more to August of 1921, with the Hay Springs News reporting the preliminary sighting in 1922. A subsequent record emerged in 1923 within the Omaha World-Herald, wherein J.A. Johnson, along side two other guys, mentioned their come across with the creature within the course of a tenting adventure along the lake's shore. They depicted the creature as just like an alligator, whole with a horn corresponding to that of a rhinoceros. Upon recognizing the men, the monster is alleged to have thrashed its tail violently earlier than submerging itself under the water's floor.

While a few skepticism surrounds the existence of the Alkali Lake Monster, particularly in mild of the truth that John G. Maher, regarded for his penchant for fabricating tall recollections, changed into related to the Hay Springs News at the time of the preliminary file, the legend persists. Furthermore, historical facts suggests that a immoderate drought hit the Alkali Lake area among 1889 and 1890, substantially lowering the lake's water stages. This has led a few to impeach the feasibility of a 40-foot monster dwelling inside the lake virtually 3 many years later.

Regardless of the skepticism, the legend of the Alkali Lake Monster keeps to fascinate human beings with an hobby in cryptozoology, closing a outstanding problem matter of discussion and exploration. Today, despite the fact that sightings of the monster have come to be uncommon, its legend endures, captivating the imaginations of folks who are attracted to the mysteries of the unexplored and the unexplained.

Nevada: Water Babies of Pyramid Lake

Nestled in the beautiful landscapes of Nevada, Pyramid Lake serves as a beacon for photographers and anglers at some point of the globe, interested in its breathtaking scenery and full-size fishing opportunities. However, underneath its serene ground, nearby lore speaks of an eerie and darkish presence that haunts the lake's depths.

According to the neighborhood Paiute legends, the lake is inhabited via using malevolent entities known as water toddlers. These vengeful spirits are said to have originated from a worrying beyond, in which the Paiute people, in an try to maintain the strength and fitness of their tribe, allegedly stable away toddlers who were both untimely or born with deformities, leaving them to perish in the bloodless waters of Pyramid Lake. It is concept that the spirits of those harmless souls converted into water toddlers, searching out to precise their revenge on every person who dares to mission into their aquatic realm.

The water babies are described as nefarious creatures that lurk below the water's floor, patiently searching out the proper 2d to claim their patients and drag them right all the way down to a watery grave. Tales in their malevolence are not simply restricted to folklore, as there have been numerous money owed of mysterious disappearances and inexplicable sounds echoing in some unspecified time in the future of the lake, particularly in the course of the springtime.

Several unique variations of this tale have circulated all through the years, but the most famous and chilling one tells of the souls of the discarded babies, their anger and thirst for vengeance turning them into the menacing water babies that dangle-

out the lake to this contemporary. The cries of these tormented spirits are said to resonate across the lake, serving as an ominous warning to those who dare to tread near their location.

Despite the bone-chilling testimonies that surround Pyramid Lake, its appeal continues to captivate the creativeness of many, drawing them to find out its mysteries and possibly even encounter the frightening water infants themselves. Whether those recollections are mere fables concocted to push back the curious or a true testament to the lake's haunted beyond stays a thriller, one which provides an thrilling and ominous duration to this in any other case picturesque area.

New Hampshire: Wood Devils of Coos County

The far flung and rugged terrain of Coos County in New Hampshire is a excellent habitat for a cryptid to stay hidden from the human eye. Known for its dense forests and the chilling temperatures that pervade the place, Coos County has lengthy been related to stories of the Wood Devils - elusive, mysterious creatures which can be said to roam the woods close to the Canadian border.

The Wood Devils are defined as towering over seven toes in top, with a slender and emaciated appearance that lets in them to effects camouflage themselves in the back of timber and one-of-a-kind foliage. Their our our bodies are covered in a thick coat of grayish fur, which affords to their capacity to mixture into the shadows of the wooded region. While their visage may be fearsome, reviews from locals and those who have claimed to look those creatures united states of america that they may be non-competitive, opting to cover from human touch as opposed to interact with it.

The legend of the Wood Devils can be traced again to the early twentieth century, with sightings peaking within the 1930s. These creatures are said to be similar to Bigfoot, every other cryptid recognized for its elusive nature. However, there are stark variations between the 2, with the Wood Devils being described as a extremely good deal thinner and further agile than their Bigfoot opposite numbers. The grayish fur that covers their our bodies additionally distinguishes them from the darker colours typically related to Bigfoot sightings.

The name "Wood Devils" have become given to these creatures with the aid of manner of locals who would in all likelihood often pay attention their eerie, coronary coronary coronary heart-wrenching screams emanating from the wooded area. These sounds, combined with the occasional footprint or fleeting glimpse of a tall, thin discern darting in the returned of a tree, have delivered approximately the myth of these cryptids embedding itself in the folklore of New Hampshire.

The story of Will, an beginner photographer from northern New Hampshire, gives another layer to the legend of the Wood Devils. In his account, he describes a close to stumble upon with the shape of creatures at the same time as seeking to photo moose near the Pontook Reservoir. His description of the Wood Devil he noticed is everyday with one-of-a-type reviews, together with credibility to the existence of these cryptids.

Despite the lack of concrete proof, the legend of the Wood Devils of Coos County remains a topic of fascination and intrigue for every locals and visitors alike. The dense forests and a long manner flung vicinity of the area make it a pinnacle location for a cryptid to live hidden, and the rich records of sightings and folklore surrounding the ones creatures handiest presents to the thriller. Whether they may be a first-rate species, a sub-species of Bigfoot, or a few element without a doubt considered considered one of a kind stays to be seen. However, one problem is for exceptional - in case you find your self hiking within the mountains of New Hampshire, preserve your eyes peeled, for you will probable in reality have a run-in with the elusive Wood Devils of Coos County.

New Jersey: Jersey Devil

At the Pine Barrens of Southern New Jersey, a close-by cryptid legend referred to as the Jersey Devil has been terrorizing humans in the place because the 1700s.

Origins

According to the legend, this monster changed into the 13th and an unwanted toddler of Mother Leeds – a South Jersey lady inside the colonial generation. While despite the fact that in her womb, Leeds gave the kid to Satan. As a prevent end result, some say that the child grow to be born deformed. Others say it changed into born ordinary but took some bizarre trends later, e.G., giant horse-like head, winged shoulders, elongated frame, and so on. The child was imprisoned till it escaped, both up the chimney or through the cellar door.

Appearance

Multiple encounters with the Cryptid over the years have left an in depth portrait of the Jersey Devil: a 7-foot tall monster with the pinnacle of a horse, large bat-like wings, a reptilian body, prolonged, spindly legs, and palms with claws. It has pink-colored eyes that seem to glow like embers.

Sightings

Over the years, hundreds of stories of encounters and sightings have been mentioned. There have even been rewards furnished for the monster's seize. The Jersey Devil has been accused of many things, from unfavourable climate, cattle lack of lifestyles, crop failure, and supposedly inflicting nearby streams to boil.

Chapter 7: The First Encounter

Around the start of the nineteenth century, many New Jersey residents cautioned sightings of the creature. The first cited sighting grow to be from Joseph Bonaparte – the previous King of Spain – in 1812, who claimed to peer the Cryptid even as he became searching.

The second exquisite sighting took place in 1909 whilst Commodore Stephen Decatur practiced cannonball taking pictures along his colleagues inside the Navy. According to his claims, one cannonball hit the Jersey Devil however did not harm the Cryptid. After his record to Navy officers, the locals had a huge panic. Following this incident, extra than 1,000 reviews of creature sightings/encounters were made through manner of way of residents.

Sighting in Salem City

In 1927, a cab reason strain said each different sighting of the Jersey Devil. At nighttime in Salem City, he pulled over for a flat tire and shortly heard terrifying screeches. Suddenly, the monster regarded from the woods and attacked him. The cab riding stress stated the incident to the police, however they could not determine out what creature had attacked him.

Devil at the Road

In 1972, Mary Ritzer Christianson changed into using one night time down Greentree Road at the same time as she observed a creature crossing the road inside the decrease lower back of her in her rearview replicate; her description of the creature matched that of the Jersey Devil.

More Sightings in the 1980s

Several greater incidents came about inside the Nineteen Eighties. One exceptional document was from an Asbury Park Press reporter who claimed that he encountered the monster a few years in the past. Another great sighting happened inside the past due 1980s even as a group of pals using bikes and ATVs across the Pine Barrens claimed to pay hobby the terrifying screams of an unknown creature.

Recent Stories

David Black said a sighting in 2015. According to him, he noticed the creature on Route 9 in Egg Harbor Township. He first idea it turn out to be a llama but quick witnessed it spreading its wings and taking flight. Black even claimed to capture a photograph of the Cryptid as evidence of his come across. While many humans have disproven his claims and picture, Emily Martin, who moreover claims to have seen the creature, shared a video that she stated turned into recorded at a place about a few miles from Black's come upon. The video portrays a similar creature.

Whether or not the testimonies of the Jersey Devil are actual, the creature has really made its manner into New Jersey lifestyle. The legend moreover inspired the decision of the neighborhood hockey organisation, "The New Jersey Devils," and loads of NJ residents though bear in mind that the tales are proper.

New Mexico: La Llorona

The sinister photograph of La Llorona, or "The Wailing Woman," a spectral decide cradling a apparently harmless little one in her spectral arms, is a deeply rooted and pervasive legend that has haunted the cultural tapestry of New Mexico for generations.

This time-venerated human beings tale, steeped in tragedy and horror, has severa iterations and deviations which have emerged through the years, with the maximum widely stated and conventional model tracing its origins to a tale concerning a stunningly beautiful lady named Maria. In this story, Maria, blessed with airy splendor, captures the coronary heart of a rich and handsome rancher, their union culminating in the shipping of children. The preliminary bliss of their marriage, but, step by step offers way to discord and estrangement, because the rancher's interest is inexorably drawn far from Maria, focusing mostly on their offspring.

This emotional chasm reaches its zenith on the same time as Maria's eyes are inadvertently attracted to the sight of her husband inside the business enterprise of each different woman, a revelation that ignites a firestorm of rage inner her. Consumed through her anger and looking for a twisted form of retribution, Maria makes the fateful selection to forged her youngsters into the swirling depths of the river. This impulsive act is swiftly observed through a maelstrom of regret, compelling her to take her personal existence inside the equal river that claimed her youngsters.

That very night time time, an eerie specter garbed in the burial get dressed that once enshrouded Maria—an prolonged, flowing white gown—become seen traversing the riverbank, her mournful wails permeating the night time air as she engaged in a apparently limitless quest for her out of area youngsters. The plaintive cry of "¿Dónde están mis hijos?" or "Where are my youngsters?" reverberated through the village, placing fear into the hearts of all who heard it.

In one-of-a-type incarnations of the tale, the narrative takes on a greater sinister flip, portray Maria as a malevolent decide from the onset. In those variations, she intentionally drowns her kids as a way of freeing herself to be with the person she desires. However, this man spurns her advances, propelling her into the abyss of despair and main her to take her very very own life. Upon her loss of lifestyles, the gates of heaven are closed to her, as she is bereft of her youngsters. This eternal separation from her offspring condemns her to roam the earthly aircraft in perpetual are seeking out of her drowned progeny.

In the gadget of her countless quest, La Llorona is said to abduct kids who each undergo a resemblance to her misplaced children or folks that showcase disobedience inside the course of their mother and father. The ominous determine of La Llorona, because of this, serves as a cautionary tale, a spectral admonition that underscores the importance of filial piety and obedience.

Over the years, the legend of La Llorona has morphed and advanced, with severa variations and interpretations emerging, each adding a very precise layer to this rich and multifaceted folklore. The essence of the story, however, remains a haunting parable that continues to resonate and instill fear inside the hearts and minds of folks who dare to delve into its darkish and twisted depths.

New York: Champy

The enigmatic creature known as Champy, or Champ, this is stated to inhabit the waters of Lake Champlain has been a fascinating state of affairs of neighborhood folklore, fascinating the imaginations of residents and vacationers alike for hundreds of years. This mythical lake monster has been said with the beneficial aid of loads of eyewitnesses, with sightings dating again to as early as 1819.

The first recorded point out of this creature become thru the Abenaki tribe, who stated it as "Tatoskok." However, the creature's mythos actually started out to take form in 1819 even as Captain Crum, sailing close to Bulwagga Bay, claimed to have visible the darkish-skinned creature. He described it as being about two hundred ft long, with 3 teeth, peeled onion-colored eyes, a crimson belt spherical its neck, and a white film superstar on its brow.

The form of said sightings escalated in 1873. One of the primary super evaluations that three hundred and sixty five days came from a railroad institution, who claimed to have visible a massive serpent adorned with glowing silver scales. Following this, Sheriff Nathan H. Mooney defined the creature as being an large beast, about 25-35 toes in length.

The creature's prominent fame in nearby folklore triggered each the states of Vermont and New York passing resolutions within the Eighties which might be seeking out to shield Champy, want to it truely exist. This prison popularity is a testomony to the creature's significance in nearby manner of life and its functionality effect on nearby tourism.

However, it is important to famend that the legend of Champy has been surrounded through way of controversy and skepticism. For example, at the identical time as it is generally claimed that Samuel de Champlain end up the number one to peer the mythical creature, his journals imply that he changed into without a doubt describing a garpike showed to him with the beneficial useful resource of indigenous guides. These massive fish, specially the Longnose Gar, are regarded for their wonderful period, silver-grey scales, and enamel-filled snouts, which may moreover have contributed to the legend of a lake monster.

Similarly, the well-known photograph taken by means of way of Sandra Mansi in 1977, which allegedly depicts Champy, has confronted scrutiny. Although no evidence of tampering modified into positioned, there remains no conclusive evidence that the photo indicates a residing creature. The picture's similarity to the notorious Loch Ness Monster photo, which modified into later uncovered as a hoax, further affords to the skepticism surrounding the authenticity of the Mansi photograph.

Despite the controversies, the legend of Champy stays an vital a part of the cultural fabric of the Lake Champlain vicinity. It has not best served as a source of intrigue and thriller but moreover as a photo of network satisfaction and a precious asset for community tourism. The creature's iconic popularity have turn out to be similarly cemented at the same time as the Vermont Expos, a minor league baseball crew, changed its call to the Vermont Lake Monsters and followed a Champ-themed mascot in 2005.

While the lifestyles of Champy may stay a topic of dialogue, the creature's importance to the folklore, life-style, and financial gadget of the Lake Champlain region is plain. The myriad sightings and memories that have emerged over the years keep to captivate the imaginations of locals and visitors alike, ensuring that the legend of Champy will persist for generations to return lower back.

North Carolina: Beast of Bladenboro

The Beast of Bladenboro, a mysterious cryptid native to Bladenboro, North Carolina, first made its presence appeared in late December 1953. This enigmatic creature is said to were responsible for the grotesque deaths of severa puppies inside the location. Reports indicated that the animals have been not honestly killed, but had moreover been drained in their blood, their heads brutally crushed, in a manner that differentiated those attacks from those generally perpetrated with the useful resource of a undergo or cougar.

Numerous locals stated seeing a creature that to start with seemed to resemble a mountain lion. However, upon closer exam, it have become apparent that the beast become considerably large than a mountain lion, with tracks that dwarfed those of any mentioned mountain lion.

The growing fear the numerous neighborhood populace introduced about prepared hunts for the creature, with the most brilliant of these hunts taking location in January 1954. The first said fulfillment in this hunt came from a nearby farmer named Luthor Davis, who claimed to have trapped a bobcat, in the end reporting it to the authorities. This led the mayor to claim that the threat of the beast have been removed.

However, this modified into not the very last said come upon with the Beast of Bladenboro. On the equal day as Davis's record, Bruce Soles claimed to have struck and killed a massive cat, expected to weigh among seventy five and 90 pounds, collectively together with his vehicle. This incident reignited the communicate approximately the actual identity of the beast. An extra, unnamed man or woman additionally claimed to have slain the beast, even though reviews of this incident were conflicting and inconclusive.

As no further assaults have been recommended thinking about the reality that 1954, it's miles considerably believed that the Beast of Bladenboro is no longer a hazard, whether or not or no longer because of its demise or some other unknown problem.

The legend of the Beast of Bladenboro has for the reason that become a large issue of nearby folklore. The creature has been defined in numerous strategies, with some placing ahead that it resembled a go through, at the same time as others maintained that it have become greater tom cat in appearance, much like a panther. Reports of the beast's period additionally numerous, with estimates beginning from the dimensions of a mountain lion to some thing some distance huge.

In addition to the conflicting descriptions of its look, the beast's modus operandi additionally defied easy categorization. The way in which it crushed the heads of its sufferers and tired their blood have become now not like the same old conduct of any acknowledged predator inside the vicinity.

Despite the fear and mystery surrounding the Beast of Bladenboro, a few residents capitalized on the feeling, with one nearby sign painter even producing bumper plates which have a take a look at "Home of the Beast of Bladenboro." In contemporary times, the legacy of the beast lives on, with the network booster institution Boost The 'Boro protecting an annual "Beast Fest," in which the cryptid serves due to the truth the mascot, a testomony to the indelible mark this enigmatic creature has left at the network of Bladenboro.

North Dakota: Thunderbird

The Thunderbird, a prominent cryptid woven deeply into the oral traditions and cultural statistics of North American Native tribes, has involved human beings for hundreds of years. This mythical creature is a image of power and power that spans for the duration of numerous cultures in the American Southwest, North Dakota, and the Southeastern United States.

Steeped in wealthy mythology, the Thunderbird is perceived as a supernatural entity this is living some of the mountain peaks, swooping all the way right down to take preserve of its prey — which consistent with some recollections, includes human beings — with its great, bold talons. Legend has it that after the Thunderbird took flight in pursuit of its prey, its powerful wings can also stir up violent thunderstorms, whilst flashes of lightning emanated from its eyes or beak to light up the sky.

When we delve into the cultures of the Pacific Northwest Coast, we discover that the Thunderbird holds a mainly terrific and revered place within the art work, songs, and oral histories of many tribes. Moreover, the Thunderbird has moreover placed its manner into the folklore of tribes living inside the American Southwest, East Coast, Great Lakes, and Great Plains regions.

In Algonquian mythology, the Thunderbird is said to govern the pinnacle global, in comparison to the underworld that is ruled through the underwater panther or the Great Horned Serpent. By flapping its wings, the Thunderbird is believed to create the sounds of thunder, on the identical time as its glowing eyes are stated to flash lightning bolts geared toward creatures of the underworld.

Algonquian-speakme tribes, which include the ones in Eastern Canada (Ontario, Quebec, and similarly east) and the Northeastern United States, further to the Iroquois peoples surrounding the Great Lakes, have a rich way of existence of Thunderbird reminiscences and motifs. In some variations of the parable, Thunderbirds are portrayed as protectors of humanity, punishing folks who stray from moral codes and profitable folks who uphold them.

For instance, the Ojibwe people receive as proper with that the Thunderbirds were created via way of the deity Nanabozho to fight the malevolent underwater spirits. The Thunderbirds migrate seasonally, arriving within the spring with other birds and departing in the fall at the same time as the harmful season of the underwater spirits concludes.

Furthermore, the Menominee tribe of Northern Wisconsin tells of a first rate floating mountain inside the western sky, domestic to the Thunderbirds who manipulate the rain and hail. These Thunderbirds are depicted as warriors who warfare the first-rate horned snakes, called Misikinubik, to prevent them from devouring humanity.

The mythology of the Thunderbird isn't always limited to Algonquian-speakme tribes, however is likewise found in Siouan-speakme peoples traditionally determined throughout the Great Lakes. The Ho-Chunk tribe, for instance, believes that someone who desires of a Thunderbird all through a solitary fast is destined to come to be a struggle chief.

Interestingly, a few researchers have counseled that indigenous Thunderbird memories can also moreover have originated from Native Americans coming across pterosaur fossils. While this hypothesis stays controversial, what's easy is the pervasive and enduring presence of the Thunderbird fable for the duration of severa Native American tribes, each of which brings its precise interpretation and importance to this mythical creature.

Chapter 8: Loveland Frogmen

The Loveland Frog, as an alternative referred to as the Loveland Frogman or Loveland Lizard, is a fantastical cryptid that has captivated the imaginations of those in and around Loveland, Ohio. Standing at an implementing four feet tall, this humanoid frog has become an important a part of Ohio folklore.

The first recorded sighting of this enigmatic creature dates lower lower back to 1955, whilst a nearby man said a amazing stumble upon. As he became riding at night time time, he got here all through three frogmen fame on the aspect of the road. Intrigued, he pulled his vehicle over and decided the ones bipedal amphibians for about three minutes, noting their placing appearance that merged the geographical regions of people and frogs.

The lore surrounding the Loveland Frog modified into similarly solidified in 1972, with a couple of sightings that delivered the cryptid into the limelight. A police officer patrolling the metropolis claimed to have seen the frogman on a bridge. In the weeks that accompanied, some other officer corroborated the primary officer's account, describing a similar creature. A close by farmer additionally joined the ranks of people who claimed to have encountered the mysterious cryptid.

As with many legends, the Loveland Frog is shrouded in diverse reminiscences and sightings that best add to its enchantment. The maximum contemporary stated sighting passed off in August 2016, when two teenagers engrossed in a sport of "Pokémon Go" claimed to have seen a massive frog on foot on its hind legs.

University of Cincinnati folklore professor Edgar Slotkin likened the Loveland Frog to one-of-a-kind iconic legends collectively with Paul Bunyan, emphasizing the creature's vicinity inside the rich tapestry of close by folklore. These memories, handed down via generations, hold to make a contribution to the enigmatic nature of the Loveland Frog.

The folklore surrounding this cryptid is severa, with some stories depicting the creature as having leathery pores and skin and a wand capable of taking pictures sparks. These specific traits similarly growth the Loveland Frog to a charming trouble for the ones intrigued via cryptids and folklore alike.

With its rich records and compelling memories, the Loveland Frog has hooked up itself as a outstanding figure in Ohio's folklore, charming the imaginations of each locals and cryptid enthusiasts from round the place.

Oklahoma: Octopus

In the serene waters of Oklahoma's freshwater lakes, whispers of an eerie creature have emerged, igniting every worry and fascination among locals and cryptid fans alike. Known because the Oklahoma Octopus, this aquatic enigma is defined as a huge creature, rivaling the size of a horse, cloaked in brownish-red, leathery pores and skin, with huge tentacles that it uses to traverse its freshwater domain.

Oklahoma's lush landscape is dotted with lakes, however the Oklahoma Octopus is stated to pick the murky depths of Lake Tenkiller, Lake Oologah, and Lake Thunderbird. These lakes, tranquil and picturesque, cover a chilling statistic - a considerably high price of unexplained drowning deaths that have left many speculating on the presence of this elusive monster.

Exploring the opportunity of an octopus inhabiting freshwater lakes throws us into uncharted waters. Cephalopods are strictly marine creatures, with their physiology mistaken for freshwater environments. The concept that an octopus could not handiest live on but thrive in such situations traumatic situations our know-how of these creatures. However, cryptozoologists and enthusiasts recommend that the Oklahoma Octopus can be a relic of historic times.

Tens of masses of lots of years in the beyond, components of Oklahoma were submerged below shallow seas - an extraordinary habitat for octopuses. Over time, it's far encouraged that the ones octopuses superior and tailor-made to the converting panorama, finally finding solace within the u . S .'s freshwater lakes.

The perception of a huge, freshwater octopus is tantalizing and has captured the creativeness of many. Shows like Animal Planet's "Lost Tapes" have similarly propagated the legend, bringing the Oklahoma Octopus into the limelight and weaving it into the fabric of cryptid folklore.

Yet, as charming as this creature may be, the query stays: How did it come to inhabit manmade lakes? Oklahoma's lakes, which include the famed Lake Tenkiller, Lake Oologah, and Lake Thunderbird, had been all created within the twentieth century. If a freshwater octopus does exist, it might have had to find out its way into these lakes placed up-creation. This perception affords a giant challenge to the legend, as there can be no evidence of octopuses, freshwater or otherwise, residing in Oklahoma's rivers and lakes.

Moreover, the lore surrounding the Oklahoma Octopus is thin, lacking the rich tapestry of indigenous legends that frequently accompany cryptids. The absence of credible Native American testimonies or ancient petroglyphs in addition weakens the case for this mysterious creature.

Despite those challenges, the appeal of the Oklahoma Octopus endures. Perhaps it's the sheer improbability of this type of creature modern that captivates us, pushing the limits of our knowledge and difficult us to find out the murky depths of the unknown.

Chapter 9: Stories

STORY 1

My call is Jack, and for years, I had navigated the dense, shadowy depths of the woods, honing my competencies as a seasoned tracker. Little did I apprehend that my facts must thrust me right right into a international a ways beyond the normal, a international of presidency experiments and huge creatures.

Assigned with the aid of the U.S. Administration, my undertaking have come to be to track down a Bigfoot that had escaped from a CIA technological expertise test. The dense wooded location regarded to close in round me as I ventured deeper, an unease settling in the pit of my belly. Hours have emerge as to days, and my unease morphed into outright worry.

The first glimpse of the Bigfoot despatched shivers down my backbone. Towering and imposing, it determined me with a keen intelligence that defied the expectations of a mere test. I pondered taking images it alive for the scientists, but as days surpassed, I found out I modified into outmatched.

The creature started out looking me, relentless in its pursuit thru the thick woods. Outsmarting it have end up my best wish, however the sinister fact unveiled itself as I delved into its origins. The scientists were no longer walking for the U.S. Government however a clandestine business enterprise with their private time table.

Betrayed and caught in a lethal recreation, I exposed the sudden revelation that the CIA had captured the Bigfoot, leaving me shaken. Determined to show the organisation, I dug deeper, revealing a extra horrifying reality. Human DNA were spliced with animals, growing hybrid beings released into the wild.

Driven via way of the urgency to prevent similarly damage, I placed myself one step in the back of the scientists. A entice changed into set, major to my seize and transportation to a mystery facility. The experiments that observed have been brutal, injecting me with materials that heightened my senses and surrounded thru the growls of the hybrid creatures.

Refusing to interrupt, I clung to the choice that someone might come for me. After what felt like an eternity, gunfire and acquainted voices shattered the silence. The Just man, my fellow rescuers from the U.S. Government, had observed me.

Emerging from the potential scarred but more potent, I vowed to discover the truth and bring the scientists to justice. The horrors of the deep woods were behind me, but my actual calling had truly started. I walked away, leaving the shadows in the lower back of, prepared to combat in opposition to the secrets and techniques that threatened us all. The journey had modified me, but in the end, I emerged as a photograph of resilience against the darkness that lurked within the coronary heart of the government.

STORY 2

It changed into a popular night time, the air complete of the lingering exhilaration of the film we had simply watched. My boyfriend and I strolled along the familiar route in the direction of my house, the recurring of our goodbyes just throughout the corner. Little did I recognise that this night time time would carve a notch into my reminiscence, etching an stumble upon that defied clarification?

As we reached the problem in which our paths generally diverged, my boyfriend and I lingered, engaged in informal communique beneath the dim glow of streetlights. The environment grow to be serene, till an abrupt shift caught my hobby. His gaze constant on some factor past me, his expression a aggregate of false impression and alarm. I discovered his line of sight, turning to face the unknown entity that had seized his hobby.

Across a small region from us, the night time time completed host to an enigma. A pitch-black parent, its moves paying homage to a seal, swayed from side to side. A palpable tension hung inside the air as we every determined this mysterious creature, apparently getting nearer with out advancing forward. My heart quickened, a take a seat down decrease lower back taking walks down my backbone as I requested my boyfriend if he perceived the same unnerving proximity. His reaction mirrored my worry – the creature, although immobile in vicinity, appeared to draw close to.

Our descriptions synchronized, a shared reality that left us grappling for reasons. The creature, a shadow inside the night time, become a source of unease and uncertainty. In the dim glow of the streetlights, we speculated approximately its nature, questioning the fact of what we had witnessed.

The fear lingering in the air propelled my boyfriend to offer to walk me the rest of the way home. The unknown grow to be too close to for consolation, and the eerie presence in the course of the world left us unnerved. Together, we navigated the path within the course of my residence, the feeling of being watched lingering until the final step.

When my boyfriend have become to move decrease decrease back home, the creature had vanished. The location changed into yet again shrouded in darkness, and no hint of the enigmatic determine remained. Questions lingered in our minds, the encounter haunting our mind lengthy after the night time had handed.

In the times that accompanied, we searched for answers. Were there others who had witnessed a comparable phenomenon? Was there a logical reason for what we had visible? Our are looking for led us proper proper into a realm of folklore, wherein recollections of mysterious creatures and unexplained phenomena painted a canvas of uncertainty.

Living inside the UK, we delved into close by legends, hoping to discover a connection to the creature that had momentarily disrupted our regular night time time. Yet, as we sifted through testimonies and whispered money owed, the truth remained elusive.

The enigma of that night time in no way in reality dissipated, leaving us with a lingering experience of wonder and trepidation. The reminiscence of the pitch-black creature, swaying within the night time like a mysterious specter, have become a shared secret among my boyfriend and me – a tale etched into the tapestry of our opinions, all of the time unanswered and unsettling.

STORY three

My wife and I had released right right into a adventure to the Smoky Mountains from our domestic in Ohio, searching for the tranquility of nature for our anniversary getaway. Unlike most, we opted for the scenic and slower-paced united states routes, paying homage to the roads I traveled in my youth. Little did we recognize that this ride must unfold into an unforgettable nightmare, a disaster etched in our recollections.

Navigating our way with a mixture of GPS and paper maps, the standard concord in our travels gave way to confusion and frustration. Errors in our direction, a rarity for us, delivered about an uncharacteristic argument that escalated right into a heated change, the phrases flowing like disjointed languages. It changed into as though we were searching at one in all a type maps, interpreting the landscape thru lenses of incomprehension.

As we ventured deeper into a good deal much much less populated and poorer street conditions in imperative Kentucky, the wooded location have become hilly and expansive. Small cities and coffee farms dotted the panorama, set in the direction of the backdrop of federal land. The anxiety in the automobile in the long run subsided as we anxiously accompanied the erratic steerage of the GPS.

Suddenly, the device chimed in with an surprising directive: "Turn left now!" My reluctance lingered, however my partner recommended it might be an unknown shortcut. The avenue we became onto, however, hinted at a few factor more ominous. Potholes big sufficient to trap a tire, overgrown scrub, and ominous vines set the level for what spread out next.

Bringing the automobile to a prevent, I wondered my partner about the selection. Before she want to reply, she iced up, staring at her phone in an unusual, nearly robotic stillness. Instinctively, I slammed the auto into contrary, seeking to retreat from the questionable path.

As we completed the opposite flip and shifted into energy, the rear of our Lincoln changed into unexpectedly lifted off the floor, leaving us immobilized. Before comprehension may also want to seize up, a big red blob charged from the woods to our right. Its unnatural velocity and agility defied logic, transferring on all fours till it stood on hind legs, its arms accomplishing in the route of the glass.

Up near, the creature's fur discovered out itself to be strands of rotten flesh emitting a putrid scent of fish and moss. Its palms, nearly human in appearance, sported prolonged, menacing claws. The face, a twisted amalgamation of human and dog features, bore bloodshot yellow eyes that leered at my partner with a predatory starvation.

In a 2d of clarity, I felt a primal urge to guard my mate. With a bellowed command, I stomped at the fuel pedal, urging the automobile to interrupt out the clutches of the ugly creature. It scratched and banged on the car, a nightmarish pursuit that incredible ceased as quick as we broke 45 miles regular with hour, careening wildly via the winding u.S.A. Roads until the lighting fixtures of a town provided salvation.

We parked in a well-lit lot, our fingers trembling as we inspected the car. Tearful and shaken, my partner described feeling a pressure in her head and a paralyzing recognition in the path of the come upon. A cracked strut and a lump of the creature's flesh dangling from the body showed the reality of our nightmarish stumble upon.

Thoroughly shaken, we discreetly unpacked our handguns, vowing to be organized for some element may come subsequent. The memory of the creature's hungry eyes and the overpowering stench lingered as we waited till the protection of daylight hours earlier than carefully resuming our adventure at the toll road, leaving the horrors of that night time time inside the returned folks.

STORY four

As teens growing up in Northern Ohio, my pal and I had been continuously interested in the mysteries of the night time time time. Near Cleveland, near Lake Erie, the woods held an attraction of creepy noises and unidentified animal sounds that fueled our teenage curiosity. It turn out to be on one such night time that we stumbled upon an enigma that might hang-out our recollections for future years.

The moon stable eerie shadows thru the timber as we ventured deeper into the woods, our senses heightened by the atypical symphony of the night time time.

That's even as we noticed it – a floating, blinking blue moderate tracing an almost hooked J-pattern inside the darkness. At first, we brushed it off as a lightning computer virus, however this slight changed into one-of-a-kind. It emitted a top notch blue glow, nearly like a immoderate-powered LED, popularity out rather closer to the backdrop of the night time.

Curiosity getting the better human beings, we approached the slight, knowledge it became only a few toes away. The absence of diverse lightning bugs and the distinctiveness of the blue hue left us puzzled. We considered the opportunity of a person with an LED mild playing a prank, however the isolated location of the woods made that seem no longer going.

As we stared in amazement, the silence have become shattered via the use of an sudden sound – a immoderate-pitched snicker that regarded to emanate from the environment. It wasn't the sound of a normal prankster or a mischievous youngster hiding inside the woods. The incongruity of the scenario dawned on us, and a take a seat back ran down our spines.

Without converting terms, a shared experience of unease propelled us into motion. Fear gripping us, we ran out of the woods as rapid as our legs may also want to convey us, the laughing lingering inside the night time air within the lower back of us.

In the aftermath of that abnormal encounter, we determined ourselves thinking about the mysterious blue mild and its accompanying laughter. The woods, as quickly as an area of intrigue, had transformed into an unsettling realm. We shared our tale with friends and family, hoping a person may offer a proof.

Days have end up weeks, and weeks into months, but the mystery remained unsolved. No one need to provide a likely cause of the floating blue light and the inexplicable giggle that had sent us fleeing into the night time time time.

Years later, the reminiscence of that bizarre night time time time nevertheless lingers, a story we recount with a combination of surprise and trepidation. The woods near Lake Erie keep their secrets and strategies, and that blue mild, placed through the mysterious laughter, remains an unsolved enigma, a testament to the inexplicable mysteries that lurk within the shadows of the night time.

STORY five

Growing up in West Texas, in which forests were scarce and mesquite wooden dominated the panorama, my young adults spread out in a small city of an awful lot less than 2,000 humans. Despite the rural placing, the metropolis limits had been properly defined, and the streets remained properly-lit at night time time, permitting community children, my brother, and me to play until nicely past sundown. The recollections of haunted homes and mysterious figures had been a part of the close by lore, but not whatever have to put together us for the weird stumble upon that spread out one Sunday night time time in July.

Kelley, a resident appeared for his creepy demeanor, resided in a rundown unmarried-big trailer that reflected FEMA trailers from the early '90s. Tall, with curly brown hair and thick Buddy Holly-esque glasses, he had turn out to be the state of affairs of severa unsettling legends in our tight-knit network. Rumors noted nocturnal dumpster diving, the consumption of worms, or maybe a traumatic incident related to a deceased kitten.

One midnight, as a friend's mother claimed, Kelley had taken a kitten that had been run over and positioned it down the the the the front of his pants. The truth that the adults had been complicit in those memories delivered an unsettling layer of credibility. Kelley's past protected a mentally handicapped older brother named Bo, who mysteriously disappeared, leaving at the back of a static green motorcycle under the carport that supposedly "never moved over again."

Despite my propensity for spreading rumors as a infant, I avoided discussing Kelley. An incident one night, playing "monster" with a pal named Alex, had instilled a true worry in me. As we transitioned between the out of doors and the alley, Alex abruptly made a grunting, oinking noise, prompting me to dismiss it as a prank.

However, a metal thud emanating from a close-by dumpster captured our hobby. Peering into the dimly lit alley, we observed a parent limping sooner or later of the overgrown lot. The unsettling feeling of the "uncanny valley" washed over me due to the fact the decide moved with jerky, claymation-like motions. Encountering Kelley, our flashlight located out distorted hands, a flannel blouse with teats seen, and a face equal to a pig's snout.

Paralyzed with the useful resource of using worry, Alex and I stood there for what felt like an eternity, in advance than Kelley emitted a weird half of of-whoop, 1/2-squeal. We bolted, leaving the again gate open. Though my dad and mom left out my account as an overactive imagination, the uncommon occasions endured. Late at night time time, whilst within the rest room, I heard sniffing and a curious "h r m" sound outside the window, intensifying my anxiety.

Kelley have become greater reclusive after the come upon, and nobody in our circle of buddies believed our tale. One day, while he worked on his roof, Kelley sniffed the air, glanced in our direction, and moved rapid again internal. The query lingered: What emerge as he? Pale with curly hair, he did no longer fit the profile of a Native American. The uncommon speech styles, dumpster diving, and avoidance cautioned some thing greater, possibly akin to the lore of skinwalkers. My community, it appeared, harbored not simply Kelley's secrets and techniques and strategies and techniques however a myriad of weird human beings with recollections ready to be unraveled.

STORY 6

I had been a park ranger for almost a decade, patrolling the superb woods that surrounded the national park. The everyday become acquainted, the rhythm of nature calming, and the a laugh of occasional encounters with natural global stored the venture interesting. But on this precise day, the tranquility of my habitual shattered when I stumbled upon some aspect that could alternate my life forever.

It began innocently sufficient. I modified into on my conventional patrol path, taking walks alongside the well-trodden paths as quickly as I discovered strange tracks inside the dirt. They had been not like some issue I had seen earlier than, neither animal nor human, and that they regarded to steer deeper into the woods, in the route of an area strictly off-limits to the public.

Curiosity mingled with a developing experience of unease as I found the ones mysterious tracks. The forest, as quickly as complete of the sounds of chirping birds and rustling leaves, grew eerily silent. The air regarded charged with an unspoken anxiety, and I could not shake the sensation that unseen eyes had been looking my each flow into.

Hours surpassed, and the tracks led me to an surprising clearing. In the coronary heart of it stood an abandoned building, a relic from a bygone era. It seemed untouched thru manner of human hands for years, forgotten via way of time. As I cautiously approached the dilapidated shape, peculiar noises emanated from internal, sending shivers down my spine.

With each creak of the door, I have to revel in the load of the unknown pressing upon me. The interior found out a unusual scene—unexpected machinery and tool cluttered the room. At the center lay a big, steel cylinder decorated with cryptic symbols and markings. As I approached, the cylinder emitted an otherworldly glow, and in advance than I must react, a powerful stress knocked me again.

Dazed and disoriented, I awakened to a modified fact. Whispers echoed in my thoughts, and I can also want to experience things past the area of human facts. In the subsequent days, the transformation have turn out to be apparent. My senses heightened, and I started out to enjoy a connection with the forest that went past the normal.

Then, one fateful night time, I encountered the creature chargeable for the enigmatic tracks. It emerged from the shadows—a humanoid determine with mottled grey pores and pores and pores and skin and eyes that glowed with an unnatural light. Its razor-sharp claws and teeth hinted on the risk it posed. I attempted to confront it, however the creature have end up quick and elusive, overpowering me and sending me crashing into the woods.

Determined to get to the lowest of the mysteries surrounding the abandoned building, I pieced collectively a grim truth. The government had carried out mystery experiments, and I had unwittingly stumbled upon their failed introduction— the creature that now haunted the woods. I knew I needed to act, to guard every the wooded area and people who wandered inside it.

Desperation led me to touch the government, however my pleas fell on deaf ears. They dismissed me as a madman, refusing to well known the existence of the abomination now roaming the woods. Alone and isolated, I confronted a constant war within the direction of a creature that defied the laws of nature.

Each passing day examined my clear up, however I refused to give up. Armed with newfound capabilities and a self-discipline born from the duty of my discovery, I remained a lone ranger inside the coronary heart of the desolate tract. The creature have become my adversary, a manifestation of presidency experiments prolonged past awry, and I couldn't rest until it turned into captured or destroyed.

The woods that when felt like a sanctuary had come to be a battleground, and due to the fact the conflict persisted, I couldn't help however surprise how prolonged I should go through this lonely fight in competition to a creature that shouldn't exist—a combat that had thrust me into the depths of the unknown.

STORY 7

Two years ago, as quickly as I become 18, an inexplicable incident opened up in the route of our circle of relatives's annual summer season retreat to our camp in Dedham Ellsworth, Maine. Nestled about 3 hours away from our domestic, the camp, a log cabin overlooking a serene lake, held a lifetime of memories for me. However, that fateful night time time time might also want to etch an unsettling revel in into my reminiscence.

It became the useless of night time, and absolutely everyone inside the camp, which incorporates my brothers and mother and father, had retired to mattress. The most effective supply of illumination became the porch slight casting a feeble glow that slightly reached past the porch itself. Engrossed in a overdue-night time TV show, I even have turn out to be interrupted by way of a noise emanating from the kitchen. Realizing that the puppies had to skip outdoor, I took my brother's pit bulls and my tiny affenpinscher named Alfie with me.

The ritual of freeing the dogs into the yard changed into regular, a nightly prevalence in some unspecified time within the future of our stay. This time, however, have become one-of-a-type. The pitch-black darkness of the Maine woods enveloped the entirety past the porch mild's benefit, making it difficult to keep a watch on the dogs. My short-time period distraction by using the usage of manner of the sight of a loon on the lake allowed the dogs to roam freely.

When I refocused my interest, I located the pit bulls fixated on some thing inside the woods. Squinting into the darkness, I couldn't determine what had captured their attention. The popularity struck me that Alfie changed into nowhere in sight. Panic set in as I known as out for her, hearing great gentle whimpering from the course wherein the pit bulls were searching.

Fear gnawing at me, I took tentative steps toward the sound, fearing that she is probably trapped or injured. Just then, I felt motion in the back of me. Whirling spherical, I observed Alfie right at my feet, silently accompanying me. Confused and concerned, I wondered why the pit bulls had been growling and advancing, their interest despite the fact that constant at the woods.

Picking up Alfie, I started to decrease returned away slowly, trusting the instincts of the puppies who regarded to revel in danger past my information. As I grew to grow to be to retreat, a bone-chilling 2d seized me. From the depths of the woods, a distorted voice known as out Alfie's name. It appeared like an eerie mimicry of my own voice, a unsightly strive that despatched shivers down my spine. The voice wailed, and in sheer terror, I fled interior with the dogs, leaving the mysterious entity in the darkness.

Confused and shaken, I tried to make revel in of the come upon. Our camp, perched on the threshold of the lake, regarded isolated, with the closest family neighbor living at the least half of a mile in the opposite direction of the unsettling incident. What had referred to as out to Alfie, distorting its voice on this sort of haunting manner? The thriller lingered, and the as soon as-familiar woods of Maine now held an air of mystery of unease that might all the time adjust my perception of our own family retreat.

STORY eight

Between 1986 and 1989, an eerie tale spread out in the southern outskirts of Porto Alegre, Brazil, especially in the some distance off area of Lami in which my uncles and cousins lived. My now-deceased uncle ran a small "boteco," a Brazilian restaurant, in this sparsely populated location dominated thru pastures and century-antique bushes. The nearest neighbor have emerge as over a kilometer away, and lifestyles in Lami unfold out in the direction of the backdrop of a top notch and quiet panorama.

Rumors circulated a number of the locals at some stage in that component about a super creature, a hybrid with the body of a man and the pinnacle of an animal, roaming the location. Whispers mentioned its attacks on every animals and people, with tales of close to misses and misplaced livestock becoming a part of the unsettling lore. My aunt, pragmatic and skeptical, disregarded those reminiscences as mere "improvements of the people."

However, one night time, her skepticism turn out to be examined in a way she might also additionally want to in no manner have anticipated. Returning home on a bus with in reality one in every of my cousins, the darkness of the night time time enveloping them, they have got grow to be part of an incident that would ship shivers down the backbone of each person on board.

With tons less than ten passengers, a few factor big emerged from the wood and collided with the proper side of the bus, causing the reason pressure to lose control. The abrupt stop left the bus partially lodged in a ditch. Though there were no accidents, fear hung thick inside the air. The using pressure, indignant and cursing, left the bus to assess the harm and contact for help.

In the unsettling darkness, a bizarre grunt pierced the air, great from any animal or human sound. Panic ensued as passengers crowded close to the motive pressure's seat, anxiously looking ahead to his cross lower back. Minutes later, a powerful knock echoed all yet again, prompting a frantic rush to the other aspect of the bus. Cries and prayers crammed the air because the car emerge as violently shaken, threatening to overturn at any 2d.

Peering thru the windows, passengers glimpsed a big determine outside, its abilities obscured thru way of the night time. Suddenly, the chaos ceased, and an insufferable stench invaded the bus, signaling the appearance of an unseen terror. My aunt, terrified, clutched her son close to due to the fact the creature made its manner in the direction of the open door.

As the passengers huddled inside the again, a unsightly decide entered the bus—a big, naked guy with darkish pores and pores and skin, a massive goat's head wearing big horns, and yellow eyes. The creature paused, huffing angrily, earlier than retreating into the night, leaving the traumatized passengers in shock and disbelief.

Unbridled crying filled the bus till a few different arrived, summoned to help the stranded car. The 2nd riding pressure, knowledgeable of the weird occasions, were given right right down to discover the missing driving pressure. Meanwhile, the Military Brigade became activated, and ambulances, professionals, and media descended upon the scene. Newspapers protected the incident for months, featuring articles and interviews with the terrified sufferers.

Despite are searching for parties scouring the location for strains of the creature, no evidence end up decided. Theories abounded, but the thriller continued. Today, the vicinity has converted, turning into populated and hugely distinct from the time on the equal time because the unexplainable occasions transpired. The creature that attacked the bus stays misplaced to facts, leaving an unsettling enigma inside the annals of Lami's past.

STORY 9

My call is Jack, a seasoned hunter with an insatiable thirst for the amusing of the quest. The forest had end up my 2nd home, each day revel in a quest for the following massive prize. But this time, my points of interest were set on the final trophy – a first-rate deer that had eluded me for years. The exhilaration coursed through my veins as I delved deeper into the heart of the wasteland.

As the sun dipped lower in the sky, casting extended, ominous shadows a number of the towering wood, an unsettling feeling crept over me. The normal sounds of the wooded area—the rustling of leaves and the chattering of birds—have been changed with the useful resource of an oppressive silence. It changed into as though the very essence of the woods held its breath, looking my every circulate.

Determined to conquer my fears, I pressed on. However, with every passing hour, the line amongst truth and nightmare started out to blur. The wooden appeared to morph and contort, their branches attaining out like gnarled fingers, trying to ensnare me. Eerie whispers danced via the wind, voices that echoed my call in haunting tones. My coronary heart raced, and my as quickly as rational thoughts struggled to recognize the horrors unfolding around me.

I puzzled my sanity as the hallucinations intensified, questioning if the legends I'd heard whispered spherical campfires had been real. Could the Windigo, a legendary creature of terror, absolutely hold close-out those woods? Shaking my head to dispel the encroaching terror, I stumbled beforehand.

Suddenly, the various twisted bushes, the Windigo materialized. Its eyes burned with malevolent hunger, and its emaciated form loomed above me. But within the blink of an eye fixed fixed constant, the creature vanished into the shadows, leaving me to marvel if it have been real or surely some other figment of my tortured imagination.

My legs gave out, and I collapsed onto the woodland floor. Darkness enveloped me, and the final sensation in advance than unconsciousness claimed me changed into the chilling breath of the Windigo behind my neck.

When I wakened, the solar had risen, and the wooded region seemed reborn. The sounds of birdsong and rustling leaves crammed the air, shielding the horrors of the night time time earlier than. Struggling to my feet, I felt an unsettling amnesia, no longer capable of keep in thoughts the statistics of my come upon with the legendary creature.

As I made my manner once more to civilization, the ordeal of the preceding night time felt like a much off, fading memory. Yet, an indescribable fear lingered inner me. Something evil had brushed in the direction of my soul in the ones dark woods, and the recollections remained locked away, hidden even from myself. Perhaps some mysteries have been imagined to be forgotten.

Returning home, my quest for the elusive deer remained unfulfilled. Something had changed inside me—a lingering reminder that there are inexplicable forces in this international, forces that defy comprehension. The wooded location, as quick as a place of solace and triumph, now held a darker enchantment, a realm where the street among truth and delusion may additionally additionally want to blur, and the unknown may need to forever cling-out the recesses of 1's thoughts.

STORY 10

It emerge as a crisp morning in Tuskegee, Alabama, as we launched into our duck hunting journey inside the Tuskegee National Forest. The woods enveloped us, the scent of damp earth and the remote quacking of ducks such as to the anticipation inside the air. As we strolled via the dense foliage, our steps softened by the usage of the usage of the carpet of fallen leaves, we paused for a fast relaxation.

Taking a 2nd to entice our breath, I gazed up at the towering timber, their branches stretching toward the sky. That's as soon as I observed it—a fleeting flash, slightly registering in my peripheral imaginative and prescient. My interest piqued, but we had a vacation spot to obtain—the swamp wherein our duck looking day revel in would possibly unfold.

Arriving on the swamp, our attention modified into right away attracted to the left. There, at the beaver dam, some thing out of the everyday caught our eyes. A tall, wide, and totally black creature walked all through the dam, moving upright on legs. Its silhouette, closer to the backdrop of the swampy panorama, come to be every implementing and mysterious. It regarded no longer to famend our presence, focused on its moved rapid adventure into the depths of the woods.

Strangely, it regarded as despite the fact that the creature carried a large object in a unmarried hand, inclusive of an extra layer of enigma to the come across. We exchanged glances, a silent statistics passing among us that some thing we had in truth witnessed modified into past our realm of records. The forest fell into an uneasy stillness, as although shielding its breath.

Time seemed to freeze as we absorbed the sight in advance than us. The creature moved with cause, disappearing into the shadows of the woods. A loosen up ran down my backbone, and the air have end up thick with an unspoken tension. Sensing that the looking day trip had taken an sudden flip, we silently agreed that it have come to be time to call it an afternoon.

Our footsteps echoed thru the quiet woods as we retraced our path, leaving at the back of the mysterious swamp. The hunt had come to an abrupt quit, leaving us with an eerie story that might linger in our recollections. We attempted to rationalize what we had visible, thinking if our senses had finished hints on us or if the wooded area harbored secrets and techniques beyond our comprehension.

As we emerged from the woods into the fading daylight, Tuskegee National Forest stood serene and detached, its secrets and techniques and strategies hidden within the depths of the wooden. We may also moreover in no manner completely apprehend the tall, black creature that crossed our direction that day, sporting an enigma in a unmarried hand and disappearing into the shadows. The woods held its secrets and techniques near, leaving us with a haunting uncertainty that would all the time linger within the echoes of our duck looking excursion in Tuskegee, Alabama.

www.ingramcontent.com/pod-product-compliance
Lightning Source LLC
Chambersburg PA
CBHW051726020426
42333CB00014B/1168